Vitamin C - Update on Current Uses and Functions

Edited by Jean Guy LeBlanc

Published in London, United Kingdom

IntechOpen

Supporting open minds since 2005

Vitamin C - Update on Current Uses and Functions
http://dx.doi.org/10.5772/intechopen.73793
Edited by Jean Guy LeBlanc

Contributors
Fadime Eryılmaz Pehlivan, Nermin Mohammed Ahmed Yussif, Philippe Humbert, Jean Guy LeBlanc

Notice
Statements and opinions expressed in the chapters are these of the individual contributors and not necessarily those of the editors or publisher. No responsibility is accepted for the accuracy of information contained in the published chapters. The publisher assumes no responsibility for any damage or injury to persons or property arising out of the use of any materials, instructions, methods or ideas contained in the book.

First published in London, United Kingdom, 2019 by IntechOpen
IntechOpen is the global imprint of INTECHOPEN LIMITED, registered in England and Wales, registration number: 11086078, The Shard, 25th floor, 32 London Bridge Street
London, SE19SG – United Kingdom
Printed in Croatia

British Library Cataloguing-in-Publication Data
A catalogue record for this book is available from the British Library

Additional hard copies can be obtained from orders@intechopen.com

Vitamin C - Update on Current Uses and Functions
Edited by Jean Guy LeBlanc
p. cm.
Print ISBN 978-1-78923-895-2
Online ISBN 978-1-78923-896-9
eBook (PDF) ISBN 978-1-83880-644-6

We are IntechOpen,
the world's leading publisher of
Open Access books
Built by scientists, for scientists

4,100+

Open access books available

116,000+

International authors and editors

120M+

Downloads

Our authors are among the

151

Countries delivered to

Top 1%

most cited scientists

12.2%

Contributors from top 500 universities

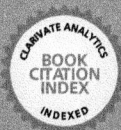

Interested in publishing with us?
Contact book.department@intechopen.com

Numbers displayed above are based on latest data collected.
For more information visit www.intechopen.com

Meet the editor

Dr. Jean Guy LeBlanc has a PhD in Biochemistry (U. Nacional de Tucuman, Argentina) and an MSc and a BSc in Biochemistry (U. Moncton, Canada). He is a principal researcher (CERELA-CON-ICET, Argentina) and postgraduate teacher (Faculty of Biochemistry, Chemistry and Pharmacy, U. Nacional Tucumán, Argentina). His principal areas of study include the use of lactic acid bacteria to increase bioactive compounds (vitamins, digestive enzymes, antioxidants, etc.) for the fermentation of foods or as biopharmaceuticals to treat and prevent vitamin deficiencies, inflammatory diseases, and some types of cancer. He has edited 5 books, published 91 peer-reviewed articles and 28 book chapters, and participated in 135 works in scientific meetings.

Contents

Preface

Vitamin C, or ascorbic acid, is mainly present in fruits and vegetables. The consumption of such foods is important since the human body does not have the ability to produce this essential micronutrient. Because it is water soluble, it can also easily be lost in cooking and long-term storage. Even though the role of vitamin C has been known since the early 1930s, only recently have researchers been actively studying and demonstrating its role and function in the treatment and prevention of many diseases. These studies will be the key to providing the scientific basis that explains why this simple but important vitamin possesses such a wide range of positive biological activities.

Authors from different countries (Argentina, Egypt, France, and Turkey) have written four original chapters relevant to vitamin C and many topics are discussed including its sources, chemical structure, metabolism, bioavailability, storage, and its many functions. Some of the latter include the role of vitamin C in the immune system and inflammation, its antioxidant properties, its depigmenting effect, its role in collagen production, and its function in metal and drug absorption, lipid metabolism, bone formation, stress control, and antimicrobial properties, among others. Its many roles in cancer, cardiovascular disease, age-related macular degeneration, and common cold prevention are also well discussed. An overview of its role in skin health and its different mechanisms of actions is described in detail. There is also an interesting chapter that shows the role of vitamin C as an epigenetic regulator because it is an important mediator between genome and environment, as it participates in the demethylation of DNA and histones, epigenome, which could help understand the role of this vitamin in cancer treatments.

Jean Guy LeBlanc
CERELA-CONICET, San Miguel de Tucuman, Argentina

Section 1

Introduction

Chapter 1

Introductory Chapter: Vitamin C

Jean Guy LeBlanc

1. Vitamins

The word vitamin was originally coined to describe amines that are essential for life. It is now known that although not all vitamins are amines, there are organic micronutrients that mean that they must be consumed in small quantities for adequate growth and are required in numerous metabolic reactions to maintain homeostasis. There are 13 vitamins that are recognized by all researchers, and these can be classified as either being soluble in fats (fat soluble) (including vitamins A (retinols and carotenoids), D (cholecalciferol), E (tocopherols and tocotrienols), and K (quinones)) or soluble in water (water soluble) (including vitamin C (ascorbic acid) and the B group vitamins). B group vitamins include the following: vitamin B1 (thiamine), vitamin B2 (riboflavin), vitamin B3 (niacin), vitamin B5 (pantothenic acid), vitamin B6 (pyridoxine), vitamin B7 (biotin), vitamin B9 (folic acid or folate), and vitamin B12 (cobalamins).

2. Vitamin deficiencies

Although all 13 vitamins are present in a wide variety of foods, deficiencies are still very common in all parts of the world. There is no magic food that contains all the vitamins; the only way to avoid deficiencies is to consume a variety of foods, which are the bases of all the nutritional guidelines, or consume dietary supplements. In addition to malnutrition, certain diseases and treatments have been shown to affect vitamin absorption or bioavailability. Furthermore, pregnant women and children have a greater need for vitamins because of their increased metabolism during cell replication.

3. Vitamin C

Vitamin C, also known as ascorbic acid, is mainly present in fruits and vegetables; citrus fruits, tomatoes and potatoes are the principal exogenous source of this vitamin. The consumption of such foods is important since the human body does not have the ability to produce this essential micronutrient. Because it is water soluble, it can easily be lost by cooking and long-term storage; fortunately, most fruits and vegetables that contain large amounts of vitamin C are consumed fresh without cooking. However, it is well known that most people do not consume the recommended five servings of fruits and vegetables that would be necessary for them to fulfill their daily recommended intake of vitamin C that is around 200 mg. Because of this problem, it is now common that ascorbic acid be used as a dietary supplement, which can easily be added to foods or consumed directly in capsules or part of multivitamin preparations.

Even though it is almost unheard of that people still can be affected by scurvy today, which is directly caused by vitamin C deficiency, its early symptoms are very common. These include fatigue, inflammatory problems (especially of the gums), depression, joint pain, and anemia. Besides an inadequate ingestion of the vitamin, other causes have been linked to vitamin C deficiency such as smoking (direct and passive), malnutrition (inadequate ingestion of eating unbalanced diets), certain drugs, and malabsorption caused by certain diseases.

The consumption of vitamin C supplements are nowadays very common, not only to prevent deficiencies but also to ensure the wide range of beneficial health effects that have been reported to be associated with the consumption of this vitamin. These include having an active role in immunity, reason for which is that many consume ascorbic acid when they have a common cold, and also because it has been stated that it can play a role in cancer (in its prevention and treatment), cardiovascular diseases, and age-related diseases such a cataracts, among others. Vitamin C is also the most commonly used antioxidant substance in foods because of its safety. This property has made it the object of numerous studies where it is used as adjunctive treatments in many bacterial and virus infections and cancer treatments, another reason that has made it the vitamin of choice by consumers to improve their general health.

4. Conclusions

Even though the role of vitamin C has been known since the early 1930s and a series of interesting studies have been performed in the 1970s, only recently researchers have been actively studying and demonstrating its role and function in the treatment and prevention of many diseases. These studies will be the key to providing the scientific basis that explains why this simple but important vitamin possesses such a wide range of positive biological activities.

Acknowledgements

The authors would like to thank the Consejo Nacional de Investigaciones Científicas y Técnicas and the Agencia Nacional de Promoción Científica y Tecnológica for their financial support.

Conflict of interest

No conflict of interests exists with the publication of this chapter.

Author details

Jean Guy LeBlanc
CERELA-CONICET, San Miguel de Tucumán, Argentina

*Address all correspondence to: leblanc@cerela.org.ar; leblancjeanguy@gmail.com

IntechOpen

Section 2

Roles of Vitamin C

Chapter 2

Vitamin C

Nermin M. Yussif

Abstract

Vitamin C or ascorbic acid is one of the most common and essential vitamins. Due to its protective role, the supplementation of vitamin C becomes a must especially during the higher pollution levels. A day after day, scientists and researchers discover new functions for vitamin C. It was and still one of the cheapest treatment modalities that could preserve and protect human beings from infections, toxification, autoimmune diseases and cancer development. The role of vitamin C in providing better esthetics exhibits great importance. Its role as anti-aging agent preserves skin color and texture. Although it is not naturally synthesized in our bodies, our food is entirely rich of it.

Keywords: vitamin C, ascorbic acid, chemical structure, bioavailability, functions, different routes of administration and needed dosage

1. Introduction

In 1970, Pauling [1] stated that "Ascorbic acid is an essential food for human beings. People who receive no ascorbic acid (vitamin C) become sick and die."

2. Vitamin C source

Vitamin C is produced only in non-humans as primate species, guinea pigs, fishes and birds [2]. Although most of the animals have the ability to synthesis their needs of vitamin C, humans suffer from mutation in the DNA coding of gulonolactone oxidase which is the main enzyme responsible for ascorbic acid synthesis [3]. Due to this mutation, the external supplement of vitamin C becomes a must [4].

The main source of vitamin C for human beings is mainly found in fruits and vegetables. Citrus fruits and other types are particularly rich sources of vitamin C as; cantaloupe, water melon, berries, pineapple, strawberries, cherries, kiwi fruits, mangoes, and tomatoes. Furthermore, vegetables are considered the main source of vitamin C due to its higher content and availability for longer period throughout the year such as cabbage, broccoli, Brussels sprouts, bean sprouts, cauliflower, mustard greens, peppers, peas and potatoes [5].

3. Chemical structure of ascorbic acid

Although vitamin C is the generic name of L-ascorbic acid, it has many other chemical names as ascorbate and antiscorbutic vitamin. L-Ascorbic acid molecule is formed of asymmetrical six-carbon atoms (C6H8O6) which is structurally related to glucose (**Figure 1**) [7, 8]. Its molecular weight is 176 with a melting point of

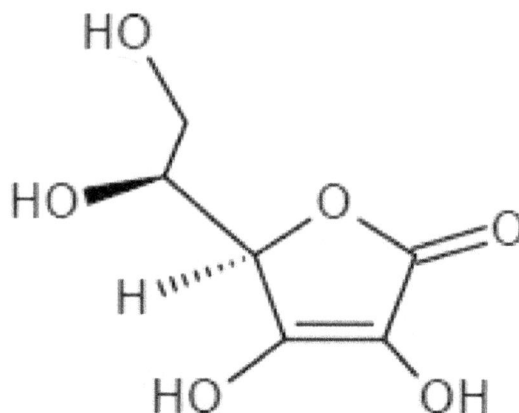

Figure 1.
Chemical structure of vitamin C [6].

190–192°C (with decomposition) and shows a density of approximately 1.65 g/cm³. L-Ascorbic acid is freely soluble in water (300 g/L at 20°C), difficult in alcohol (20 g/L at 20°C) and insoluble in chloroform, ether and benzene. It forms a clear colorless to slightly yellow solution. It has two pKa values: 4.2 and 11.6. The pH of a 5% (w/v) solution in water is 2.2–2.5 [9].

The chemical structure of ascorbic acid determines its physical and chemical properties. It is a weak, water soluble, unstable organic acid which can be easily oxidized or destroyed in light, aerobic condition (oxygen), high temperature, alkali, humidity, copper and heavy metals. Ascorbic acid is usually found in the form of white or slightly yellowish crystalline powder. Its crystalline form is chemically stable in dryness. However L-ascorbic acid is highly soluble in water, it shows great difficulty to be soluble in alcohol, chloroform, ether and benzene. In water, it forms clear colorless slightly yellow solution which is rapidly oxidized [8, 10].

There are many derivatives of ascorbic acid as sodium L-ascorbate (sodium ascorbate), calcium L-ascorbate (calcium ascorbate), zinc-ascorbate, 6-palmityl-L-ascorbic acid (ascorbyl palmitate) and ascorbyl monophosphate calcium sodium salt (sodium calcium ascorbyl phosphate) [128, 129].

Ascorbic acid is obtained from sodium ascorbate by cation exchange. While sodium ascorbate results from reacting methyl-D-sorbosonate (or ketogulonic acid methyl ester) with sodium carbonate. Calcium ascorbate is produced by the interaction of ascorbic acid with calcium carbonate in water and ethanol, which it is then isolated and dried. Ascorbyl palmitate is prepared by reaction of ascorbic acid with sulfuric acid followed by esterification with palmitic acid. Sodium calcium ascorbyl phosphate resulted from the reaction of ascorbic acid (alone or in combination with sodium ascorbate) with calcium hydroxide and sodium trimetaphosphate. The previous ascorbic acid derivatives have superior properties in comparison to ascorbic acid as the light resistance, skin irritation [128, 129].

4. Vitamin C metabolism

Vitamin C functions depend mainly on its main character as a reducing agent and the results of its oxidation mechanisms either reversible or irreversible [130, 131]. These reactions depend only on the pH changes and not on the presence of air or oxidizing agents [132].

Ascorbic acid undergoes a 3-step oxidation process. In the beginning, ascorbic acid can reversibly oxidize into dehydroascorbic acid on the exposure to copper, low alkaline media and heat [11].

Dehydroascorbic acid is a very short half-life (few minutes) product which can either reversibly or irreversibly oxidize in the tissues. In pH 4.0, ordinary temperatures and aqueous media, dehydroascorbic acid can be oxidized irreversibly into 2,3-diketo-L-glutonic acid (diketogulonic acid). However, the dehydroascorbic acid oxidation begins in mild acidic media (pH 4.0), it requires a neutral or alkaline media to progress more rapidly. The resultant diketogulonic acid is a stronger reducing agent, not reduced by glutathione or H2S and not an anti-ascorbutic agent. It was found that below pH 4.0, diketogulonic acid losses its reducing property. In acidic media and the presence of H_2S, dehydroascorbic acid can also reversibly change into ascorbic acid. Ascorbic acid and dehydroascorbic acid have the same anti-ascorbutic effect [133, 134].

The third oxidation product is L-threonic acid and oxalic acid which proceed only in alkaline media (pH 7–9) [11]. All reversible changes can be done in the presence of H_2S and glutathione in neutral or alkaline media. Sometimes, carbon dioxide may be the result of vitamin C oxidation at high doses [132].

In human beings, ascorbic acid is reversibly oxidized into dehydroascorbic acid, which can be reduced back to ascorbic acid or hydrolyzed to diketogulonic acid and then oxidized into oxalic acid, threonic acid, xylose, xylonic acid and lyxonic acid. Further oxidation (decomposition) may occur by the oxidizing agents in food. According to the oxidation-reduction reactions, ascorbic acid is the reduced form of vitamin C while dehydroascorbic acid is the oxidized form of vitamin C. The L-isomer of ascorbic acid is the only active form. Other isomers as D-ascorbic acid, D-isoascorbic acid and L-isoascorbic acid are present. These stereoisomers have no effect in the treatment of scurvy [128, 129].

The absorbed and the unabsorbed forms of ascorbic acid can be excreted in conjugated or non-conjugated pattern. Ascorbic acid may undergo limited conjugation with sulfate to form ascorbate-2-sulphate, which is excreted in the urine. Unchanged ascorbic acid and its metabolites are excreted in the urine. In the presence of intestinal flora, high doses of ascorbic acid (unabsorbed part) can oxidized into carbon dioxide which is the main excretory mechanism of vitamin C in guinea pigs, rats and rabbits. There exists equilibrium between ascorbic acid and dehydroascorbic acid, dependent on the redox status of the cells [12].

5. Vitamin C bioavailability

The bioavailability is a measure of the efficiency of gastrointestinal tract absorption [13].

5.1 Vitamin C absorption (active transport)

The hydrophilic nature of ascorbic acid facilitates its absorption through buccal mucosa, stomach and small intestine. Its absorption depends mainly on passive diffusion through the buccal mucosa [14].

Vitamin C absorption occurs through small intestine (distal intestine) by active transport mechanism. Sodium electrochemical gradient is the process by which active transport of ascorbic acid occurs. This process proceeds by the help of sodium vitamin C transporter type 1 (SVCT1). This is the same transporter responsible for vitamin C transport in retina. SVCT2 is responsible for transporting vitamin C into brain, lung, liver, heart and skeletal muscles [15]. The absorption process is usually inhibited by glucose [135].

The majority of ascorbate is transported by SVCT1 in epithelial cells (e.g., intestine, kidney and liver), and the remaining is transported by SVCT2 in specialized cells (e.g., brain and eye) [15, 16]. The main concentrations of vitamin C are located in brain and adrenal cells.

The oxidative products of vitamin C as dehydroascorbic acid are transported faster into cells than the pure form [17].

While the absorption of low doses (15–30 mg) is very high (up to 98%), ascorbic acid absorption decreases (50%) with larger doses (1000–1250 mg) which is commonly administrated in acute illness [136, 137].

In human blood, ascorbic acid is always found in the reduced form (ascorbic acid). It was also found that the red blood corpuscles are not permeable for ascorbic acid and also to glucose. It oxidized very slowly in blood than in plasma (no oxidation reactions occur) [132]. Its normal plasma level ranges between 50 and 100 μM according to the diet intake in healthy non-smoker individuals [137]. Increasing the plasma level and the intracellular level is not a dose dependent. Its intracellular level is higher than the plasma level. The plasma level does not increase above the normal range even by increasing the intake into 500 mg because of its excellent excretion from kidneys through urine [13].

5.2 Vitamin C distribution

Vitamin C is widely distributed in all the body tissues. Its level is high in adrenal gland, pituitary gland, and retina. Its level decreases in kidneys and muscles.

5.3 Vitamin C excretion

Vitamin C metabolites (oxalate salts) and unmetabolized vitamin C are excreted by kidneys. Few percentage of vitamin C is excreted through feces.

The urinary excretion of vitamin C is dose dependent. Less than 100 mg/day, no vitamin C was detected in urine. At 100 mg/day, 25% of its amount was excreted in urine. The latter percentage is doubled with the administration of 200 mg/day [13, 18].

At high doses, large amount of unmetabolized vitamin C is excreted. The higher doses of vitamin C intake, the higher vitamin C concentration in blood and tissues occurs. As a response for high doses, vitamin C excretion from kidneys and sweat occur. The antiviral and anti-bacterial effect of vitamin C protects skin and kidneys from infection [1]. Also in extra doses, the oxidation components were used as an anticancer effect more the vitamin C itself [138].

It was found that the excretion of ascorbic acid when administrating 400 mg ascorbic acid ranges between 30 and 50% in healthy individuals. This percent decreases in diseased patients due to higher consumption. Repeated low doses (about 200 mg) are highly recommended in diseased individuals due to theses low doses saturate the body. Extremely low dosages (90 mg/day) could result in inability of the immune system to respond to diseases as degenerative diseases. Therefore, limited renal clearance of ascorbic acid is usually detected. The plasma saturation of ascorbic acid at 70 μM (0.123 mg/dl). This level controls the excretion of the ascorbic acid through kidneys. At Higher plasma levels (above 70 μM), higher excretion levels are usually detected. The intravenous route exerts 30–70 folds of vitamin C plasma levels than the oral route [19].

The rapid excretion due to its water soluble nature limits its harmful effect and makes it totally safe product in normal doses. It also found that the upper tolerable limit (UL) is 2 g. Depending on the depletion-repletion study, it was found that the RDA is 75 mg for women and 90 mg for men. It was modified by Levine et al. in

	1–30 days (g)	11–45 years (g)	46–77 years (g)
Adrenals	0.581	0.393	0.230
Brain	0.460	—	0.110
Liver	0.149	0.135	0.064
Kidney	0.153	0.098	0.047
Lung	0.126	0.065	0.045
Heart	0.076	0.042	0.021

Table 1.
Vitamin C storage in different organs at different ages.

2001 into the administration of 90 mg to both sexes. The maximum bioavailability and absorption of vitamin C achieved at 500 mg [20].

5.4 Vitamin C storage

In 1936, Marinesco et al. detected the lower levels of ascorbic acid in others organs as pancreas, spleen and thymus gland. Plaut and Billow detected the ascorbic acid lowering not only in the organs but also in body fluids as CSF, blood and urine. They also detected this deficiency in neural diseases and alcoholism. Many reasons were thought to be the cause of vitamin C deficiency in old people. Decreased intestinal absorption and dietary deficiency are the main causes.

In human blood, ascorbic acid is always found in the reduced form (ascorbic acid). It was also found that the red blood corpuscles are not permeable for ascorbic acid and also to glucose. It oxidized very slowly in blood than in plasma (no oxidation reactions occur) [132]. Its normal plasma level ranges between 50 and 100 μM according to the diet intake in healthy non-smoker individuals [137]. Increasing the plasma level and the intracellular level is not a dose dependent. Its intracellular level is higher than the plasma level. The plasma level does not increase above the normal range even by increasing the intake into 500 mg because of its excellent excretion from kidneys through urine [13]. In 1934, Yavorsky et al. [21] analyzed the ascorbic acid amount found in the different body organs at different ages (**Table 1**).

6. Vitamin C functions

6.1 Role in immune system and inflammation

Vitamin C has an important role in the maintenance of a healthy immune system and its deficiency causes immune insufficiency and multiple infections. The ascorbic acid level is lowered in various body fluids during bacterial infections. Thus, it is commonly used as adjunctive treatment in many infectious diseases such as hepatitis, HIV, influenza and periodontal diseases [22].

Vitamin C administration modifies and enhances both the innate and adaptive immune response. It neutralizes the bacterial toxins especially endotoxins by blocking the essential signal for lipopolysaccharides (LPS) formation. On the other hand, LPS block the passage of ascorbic acid through blood brain barrier and inhibits its uptake by various cells [22].

Ascorbic acid improves the phagocytic properties and activity of various immune cells including neutrophils, natural killer cells, macrophages and lymphocytes. Vitamin C increases lymphocytes proliferation and antibody production [23, 24].

6.2 Anti-oxidant property

Oxidative stress/ROS have a main role in inflammatory diseases including periodontal diseases [25]. The ROS are classified into 3 classes; the first are reactive free radicals as oxygen related radicals (superoxide, hydroxyl radical or peroxyl radicals). The second class is reactive species but not free radicals as hypochlorous acid. The third class is radicals resulted from the reaction with ascorbic acid [26]. Antioxidants are also classified into enzymatic and non-enzymatic. The enzymatic antioxidants include catalase enzyme, thiol-containing agents (cysteine, methionine, taurine), glutathione and lipoic acid [27].

Vitamin C is one of the nutrient non-enzymatic anti-oxidants [28–30]. Its antioxidant effect is by electron donation process where vitamin C easily donates two electrons (reduction reaction) to other compounds in order to prevent its oxidation. When ascorbic acid donates the first electron, it is transformed into a free radical called ascorbyl radical (semi-dehydroascorbic acid). It is a relatively stable, unreactive free radical with unbound electron in its outer shell but it has a short life time (10–15 s). The unreactivity of this radical makes it unharmful to the surrounding cells. This process is called free radical scavenging or quenching. When it donates the second electron, it transformed into dehydroascorbic acid. Its stability may only last for few minutes [28, 31].

As a general rule, it was detected that vitamin C acts as a pro-oxidant at low doses and acts as an antioxidant in high doses. It was also detected that the level of vitamin C in the skin usually exposed to ultraviolet radiation is lower than that exposed lesser.

The antioxidant activity of vitamin C enhances the epidermal turn over, and the movement of young cells to the surface of the skin where they replace old cells [32]. The study conducted by Frank in [139] showed that RNA improved the ability of the skin cells to utilize oxygen.

Ascorbyl radical and dehydroascorbic acid are reversible agents which can easily rebound into ascorbic acid. These reversible agents can irreversibly transformed into 2,3-diketogulonic acid which is further metabolized into xylose, xylonate, lyxonate and oxalate (**Figure 2**) [34].

Vitamin C is considered as a strong anti-inflammatory agent as it inhibits many types of inflammatory mediators as tumor necrosis factor alpha [35]. This property is commonly used in the treatment of postoperative erythema formed after CO_2 laser in skin resurfacing [36]. In 1987, Halliwell [37] detected significant reduction of plasma levels of ascorbic acid in association with elevated histamine in inflammatory diseases as ulcerative colitis and rheumatoid arthritis. This was explained by the discovery of the anti-histaminic effect of vitamin C. It was also found that the higher ascorbic acid content in joints, the higher protection levels against damage which directed many physicians to use ascorbic acid in combination therapy with drugs aiming to joint protection as glucosamine [37, 38].

It was discovered that vitamin C has an efficient chemotherapeutic effect. The cytotoxic effect of vitamin C is dose and route dependent. The tumor cells are more sensitive to high intravenous (cytotoxic) levels of vitamin C than the normal ones [140]. At the administration of 10 g of intravenous vitamin C, a marked elevation of the extracellular concentration (1000 μmol/L) is detected which have a toxic effect on the cancer cells due to the action of the ascorbyl radicals (free radical species) [39]. On the contrary to the cancer cells, normal cells can compensate the damage occurred by these oxidative species [141]. It was also found that these mega doses of vitamin C are essential in other diseases as diabetes, cataracts, glaucoma, macular degeneration, atherosclerosis, stroke and heart diseases [40].

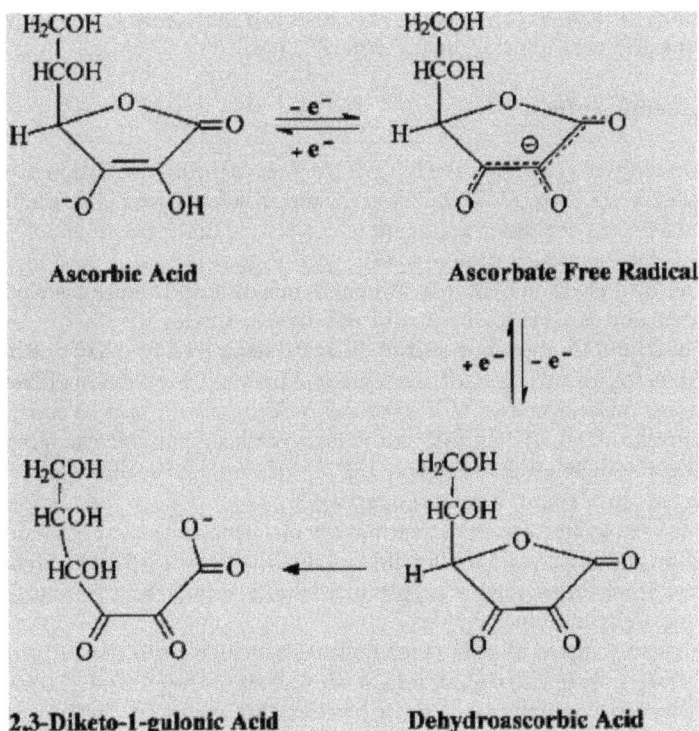

Figure 2.
Redox metabolism of ascorbic acid [33].

Vitamin C improves the immune system and its deficiency causes immune insufficiency and multiple infections. It was found that vitamin C modifies the behavior and activity of the immune cells; it also improves the phagocytic properties of neutrophils and macrophages. In addition, vitamin C increases the antibody production, concentration of antibodies and the activity of lymphocytes [41]. It was detected that the level of vitamin C in leukocytes is higher than its level in plasma because they have the ability to store it [142].

Vitamin C is commonly used as an adjunctive treatment in many infectious diseases as hepatitis, HIV, common cold and influenza. It has an important role in the antibacterial reactions performed in our body by neutralization of the bacterial toxins especially endotoxins [42]. It was found that 100 μM/L ascorbic acid can lower the bacterial replication (bacteriostatic effect) [143].

Sufficient amount of vitamin C causes blockage of the signaling essential for lipopolysaccharides (LPS) formation. It also stops the production of ROS especially reactive nitrogen species which is mainly produced during infection [42]. In bacterial infections, the level of ascorbic acid in various body fluids is lower than usual which perform further depression due to the action of LPS in blocking the passage of ascorbic acid through blood brain barrier. LPS also inhibits the uptake of various cells to ascorbic acid [143].

The anti-aging effect of vitamin C is regarding to its potent antioxidant effect, its stimulatory effect of enhancing the collagen formation, protection of the persistent collagen especially elastin against damage and finally, inhibits the cross-linking effect formed in wrinkles [144]. It was found that the amount of ascorbic acid changes with age. The younger the age, the higher the ascorbic acid level present. In 1934, Yavorsky et al. [21] analyzed the ascorbic acid amount found in the different body organs at different ages.

6.3 Depigmenting effect

It was found that the higher the ROS is, the deeper the pigmentation produced. Anti-oxidants act a great role in lowering the melanin formation [43]. Vitamin C is considered a potent depigmenting agent which is used in the treatment of various cases of skin hyperpigmentation [44–48].

It can be used as an adjunctive treatment in melasma and severe cases of hyper-pigmentation and as a treatment in mild and moderate cases [49].

Vitamin C inhibits melanogenesis in different steps via more than one mecha-nism [145]. Being an anti-oxidant, ascorbic acid prevents production of free radicals which triggers melanogenesis [50]. It reduces o-dopaquinone back to dopa, prevent-ing dopachrome of 5,6-DHICA [51] and reduces oxidized melanin changing the pigmentation from jet black to light tan [52]. Furthermore, the direct suppression of tyrosinase enzyme exhibits a great property [53].

It can be used as an adjunctive treatment in melasma and severe cases of hyper-pigmentation and as a treatment in mild and moderate cases [49]. The higher the ROS is, the deeper the pigmentation produced. Antioxidants act a great role in lowering the melanin formation [43].

Other mechanisms of blocking melanogenesis include inhibition of tyrosinase activity by interacting with copper ions at active sites of the enzyme [51, 53]. In addition, vitamin C inhibits melanocyte proliferating agents (IL-1, MSH, and PGE2) and peroxidase reactions on melanocytes [54].

In 1970, Pauling and Cameron [146] discovered that vitamin C has an efficient chemotherapeutic effect. The cytotoxic effect of vitamin C is due to the action of the ascorbyl free radical and it is dose as well as route dependent [55]. The tumor cells are more sensitive to high intravenous (cytotoxic) levels of vitamin C than the normal ones [23, 56]. A synergistic effect is detected between the intravenous vitamin C administrations accompanying the tumor cytotoxic agent in patients suffering from advanced cancer [57].

Melanoma is the most commonly treated malignant tumors using vitamin C due to the high susceptibility and sensitivity of its cells to vitamin C. It induces sodium ascorbate induced apoptosis of melisma. The lethal effect of ascorbic acid is attrib-uted to inhibiting the production of IL-18 (essential for melanoma proliferation), change the intracellular iron level [56, 58].

Besides the antioxidant role, ascorbic acid also acts as an electron donor for eight enzymes. Three of these enzymes are involved in collagen formation [34]. Other two enzymes are responsible for carnitine formation, one enzyme is responsible for epinephrine production from dopamine, and the other is responsible for the addition of the amide groups into peptide hormones. Finally, it is essential for tyrosine metabolism and melanin production. The anti-tyrosinase enzyme occurs at 0.05–0.50 mM intracellularly [31].

6.4 Collagen production

The role of vitamin C in collagen formation is well known. Vitamin C is an essential factor for the hydroxylation of proline, cofactor during collagen processing, activation of pro-collagen messenger RNA, inhibition of matrix metalloproteinases (MMPs) that are responsible for collagen fibers degradation and fibroblast activation intended for new and proper collagen formation [12, 59–61].

As regards the effect of ascorbic acid on periodontal ligament, it enhances the periodontal ligament maturation and renewal by induction of the collagen formation especially collagen III (young collagen) and keeps the balance between collagen I (mature collagen) and III for tissue maturation. It was detected that thicker periodontal ligament were detected near the CEJ and narrower ones were detected in the middle one third of the root due to the effect of vitamin C in keeping the collagen bundles, well organized and more resistant to tension. Furthermore, it also activates the fibroblast itself; proliferation, production and differentiation. By the vitamin C role in modifying the produced collagen type IV through its role as a cofactor in hydroxyproline synthesis and improving the endothelial cell vitality, its role in angiogenesis could not be forgotten. In periodontal disease, it is recommended to use an adequate well calculated dosage of vitamin C to achieve higher level of healing, minimal bleeding, higher quality of the newly formed tissues and increasing the resistance of tissues to future destruction [12, 62].

Collagen is the main component of bone matrix. Many publications had confirmed the role of vitamin C in bone formation. In postmenopausal women, higher levels of vitamin C are needed in order to reduce the incidence of osteoporosis [63, 64].

When vitamin C is used with scaffolds in tissue engineering, the sustained release of vitamin C stimulates the formation of type I collagen and alkaline phosphatase enzyme [65].

Vitamin C is a potent factor in the extracellular bone matrix proteins formation as collagen type I, osteonectin and osteocalcin. Its combination with vitamin E has an essential role in the proliferation and differentiation of the osteoblasts [66].

The amount of ascorbic acid in human body changes with age. The younger the age is, the higher the ascorbic acid level present [67].

Vitamin C enhances the collagen formation (collagen type I) and protects the persistent collagen to resist damage. Finally, it inhibits the cross-linking effect found in wrinkles [68].

The anti-oxidant property is also involved in the anti-aging effect; vitamin C plays an important role in protecting the cellular integrity as it scavenges the ROS, prevents oxidation of the cellular proteins, lipids as well as DNA and protects the cellular junctions. It also improves the tissue vasculature [69].

As regards the effect of ascorbic acid on periodontal ligament, it enhances the periodontal ligament maturation and renewal by induction of the collagen formation especially collagen III (young collagen) and keeps the balance between collagen I (mature collagen) and III for tissue maturation. Thicker periodontal ligament were detected near the cement-enamel junction (CEJ) and narrower ones were detected in the middle one third of the root due to the effect of vitamin C in keeping the collagen bundles, well organized and more resistant to tension. It also modifies the rate of fibroblast proliferation [12, 62, 70] (**Figure 3**).

6.5 Others functions

6.5.1 Vitamin C and metal absorption

Vitamin C increases the absorption of heavy metals from the intestine as iron. Vitamin C has an important role in the carnitine synthesis which is an enzyme co-factor that increases the absorption of non-haem iron in GIT. It also enhances production of reduced iron which is the preferred form for the intestinal absorption [72]. It was found that factory workers have high oxidative stress. The cause behind the latter observation was the high plasma levels of lead (73 µg of lead/dl), thiobarbituric acid (46.2%) and chloramphenicol acetyltransferase (70.3%). By the

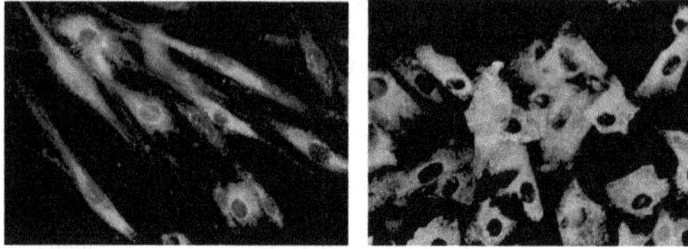

Figure 3.
(a) Human fibroblast culture and (b) human fibroblast proliferation after treated with vitamin C [71].

administration of combination dosage of vitamin C (1 g) and vitamin E (400 IU), great improvement of the general health with the plasma levels returned to its non-lead levels [73].

6.5.2 Vitamin C and drug absorption

Combination of vitamin C supplements with aspirin and opiates has a strong synergistic effect on these drugs [54, 74]. On the other hand, oral contraceptive pills increases ascorbic acid turnover and reduce level of ascorbic acid [75].

In addition, many vitamins including vitamin E (α-tocopherol), vitamin B15 (carnitine), tryptophan and folic acid require vitamin C for their absorption [76]. The combination of vitamin C and E enhance the efficiency and life span of vitamin E by providing sustained release effect and regeneration of the oxidized vitamin E [59, 77]. Thus, vitamin C deficiency is considered a cause of some of these vitamins deficiency such as folic acid deficiency [78–80].

Furthermore, marked increase of ascorbic acid turnover was reported on consumption of estrogen containing medications [81].

As a result of oxidation reaction, the production of hydrogen peroxide is enhanced by cations as iron and copper [19].

EDTA is the only permissible preservative that could be used with injectable vitamin C products [82].

Limited evidence suggests that ascorbic acid may influence the intensity and duration of action of bishydroxycoumarin.

6.5.3 Vitamin C and lipid metabolism

Although vitamin C is water soluble, it has great effects on the lipids either intra-cellularly or extracellularly. Vitamin C is an essential factor protecting the lipids of cell membrane from oxidation. It also protects the lipid bilayer of skin.

Significant control of hypertension and high cholesterol levels [147]. It was observed that vitamin C improves the lipid metabolism by inhibiting the oxidation of the unsaturated lipids and lipoproteins (scavenging effect) [83]. ROS can easily oxidize the cellular lipoproteins and the circulatory LDL and results in lipid peroxidation. By a process called radical propagation, the formed lipid peroxides can easily reacts with oxygen to finally form lipid hydroperoxides. Ascorbic acid can easily inhibit this process by reducing the ROS [84]. It was found that vitamin C decreases the oxidation of lipids and in lowering the low density lipoprotein (LDL) cholesterol [85].

Vitamin C is a cofactor in many processes in various organs including catechol-amine biosynthesis and steroidogenesis as well as lowering cholesterol and bile acids [86–88]. However, vitamin C deficiency is linked to many diseases and disorders such as fatty liver [89], hyperlipidemia [90], obesity and diabetes mellitus type II [91].

Vitamin C has an important role in decreasing the atherosclerosis and the inhibiting the thrombus formation through decreasing the platelet aggregation [148].

6.5.4 Vitamin C and bone formation

As previously mentioned, vitamin C increases the production of collagen type I and X needed for matrix formation, activation of osteoblast growth and differentiation [92]. It is also needed in order to maintain adequate bone density [93, 149]. Stimulation and higher expression of osteocalcin and osteonectin on the osteoblasts was also reported [66].

In postmenopausal women, higher levels of vitamin C are needed in order to protect against bone abnormalities as it is considered as delaying osteoporosis factor [149]. Furthermore, in scurvy, lower bone density with marked bone abnormalities commonly reported [94, 95]. In case of deficiency occurs in young individuals, bone fragility, cartilage resorption and fracture of growth plates. The detected abnormalities were attributed to reduced activity of osteocytes and chondrocytes [93, 96]. It also maintains and preserves the balance between osteoblasts and osteoclasts [97].

In order to achieve optimum proliferation of the osteoblasts and fibroblasts, 200 μg/ml is the maximum dose needed. Apoptosis occurs when exceeding such dose [39, 98].

In 2004, an in vitro study used vitamin C with scaffolds in tissue engineering in order to regenerate bone. The sustained released vitamin C stimulates the formation of type I collagen and alkaline phosphatase [65].

In 2013, Fu et al. [150] used isotonic irrigation of ascorbic acid derivative during grafting of the anterior tendon of the knee joint. Significant reduction of the inflammatory response in the surgical site due to lack of toxicity, irritation, watery consistency and potent anti-oxidant effect.

6.5.5 Vitamin C and stress (cortisone)

Vitamin C has an important role in controlling the stress [41]. In stress, the overproduction of cortisone affects the defense mechanisms [151]. This was explained by the reduction of glucose levels [152].

6.5.6 Anti-microbial properties

Although the toxic effect of high doses, it shows great benefits on the other hand. Vitamin C excretion depends mainly on kidneys and sweat. The antiviral and anti-bacterial effect of vitamin C protects skin and kidneys from infection [1]. Also in extra doses, the oxidation components were used as an anticancer effect more the vitamin C itself as mentioned before [138].

7. Vitamin C dosage

The dilemma behind vitamin C dosage has been started many years ago. Dosage calculations differ according to the medical status, aim of administration either prophylactic or curative, route of administration and patient age.

When we talk about ascorbic acid average dosage, we have to differentiate between four terms; the estimated average requirement (EAR), adequate intake level, the tolerable upper intake level (ULs) and the recommended dietary allowance (RDA). The EAR is used to calculate the RDA and the adequate intake levels. While the UL is the level below which toxic effects have not been seen [99].

Because of the poor oral bioavailability of vitamin C, toxic signs and symptoms may appear with large doses exceeding 1000 mg or utilizing more than 2 g as single dose [100].

7.1 Dosage for healthy individuals

For healthy individuals, vitamin C dosage differs according to age and sex. The average normal content of ascorbic acid in human body is about 1.5 g. Daily, our body usually consumes 3–4% (40–60 mg) of this pool. In order to keep this pool balanced, the daily oral intake should be 200–300 mg of vitamin C. The daily intake of 5 fruits or vegetables will provide a 200 mg of vitamin C which will result in 70 μM plasma level. It was also found that the uptake of vitamin C differs from tissue to another [13, 31].

As an average, the RDA of vitamin C for adult healthy men is about 90 mg and for healthy women is about 75 mg. The ULs dosage was calculated at a level of 2 g per day for healthy individuals [99, 100]. Based on gender, the RDA of both males and females should not be less than 90 mg which secure the neutrophil saturation and urinary excretion. The maximum bioavailability could be reached at 500 mg dose [155]. It was found that the oral route could produce a maximum 220 μmol/L of vitamin C plasma concentration even with high dosage administration while the intravenous route produces higher levels which could reach to 1760 μmol/L [47].

The recommended dosage also differs according to the country or health institute that recommended this dosage. The recommended dosage of vitamin C can only maintain its plasma level constant up to 5–6 h only [57].

Another suggestion said that the recommended dosage of vitamin C is 100–120 mg/day. It was discovered that dose elevation could not control developing of related diseases (Gomez-Romero et al., 2007). In 2010, the Japanese ministry of health recommended 100 mg for both men and women [153]. Older researches recommend the average suitable doses for vitamin C is 100–3000 mg/day. In 1999, Levine et al. [101] determinate the average daily range of vitamin C is 210–280 mg only by food.

The bowel tolerance dose is the dose just below which produces diarrhea. The bowel tolerance usually differs from one to another according to the medical status. It can be determined only by trail. High dosage ranging between 3 and 6 g is recommended till diarrhea occurred. The dosage then has to be decreased till the bowel balance achieved again. Such side effect (diarrhea) is useful in treatment in patients with constipation [102]. Vitamin C is well absorbed up to 500 mg/day [103].

7.2 Dosage for smokers

A higher dose of vitamin C is strongly recommended in order to compensate for the smoking hazards and neutralize for the resulted oxidative stress [104]. In 2000, Sargeant et al. [105] reported the protective effect of vitamin C administration in smokers against obstructive airway disease.

A 110 mg is the recommended daily dosage for smokers due to its antioxidant level is below normal. The lower plasma level of ant-oxidants regards to their lower consumption healthy food, higher levels of toxic products which produce oxidative stress [106, 107].

7.3 Dosage in surgery and illness

Vitamin C is an important partner in the parental nutrition especially in acute conditions (shocked surgical, trauma, burn and septic conditions) [99]. In surgery and illness, redistribution and reabsorption of vitamin C occurs. It can be assumed that increased demand in tissue results in a decrease in its plasma concentration. It was discovered that patients can tolerate much more doses of vitamin C than healthy individuals [22, 102, 108].

During acute inflammatory phase as postsurgical, trauma, sepsis, and burns, the plasma level of vitamin C significantly lowered [154]. After surgery, the blood concentrations of vitamin C decreased immediately and 7 days postoperative (44.3–17.0 mmol/ml). Also the urinary excretion of vitamin C is lowered in the first week postoperatively (3.12 and 1.94 mg/day) [155]. In severe burns, Tanaka et al. [109] used 95 g/day of vitamin C for patient weight of 60 kg in the first 24 h. Lowering the tissue edema is observed on vitamin C administration.

A 3 mg orally or 100 mg parentally is recommended in acute conditions to compensate the diseased conditions and maintain the physiologic functions [99].

In 2010, Fukushima and Yamazaki reported the intense decrease of the plasma vitamin C level in surgical complicated patients, especially surgical intensive care unit patients which leads to increase of the oxidative stress. It was found that the recommended daily doses of vitamin C (RDA) either orally or parentally are not sufficient. They recommended higher doses of vitamin C in order to compensate the large amounts consumed during illness. A 500 mg of parental vitamin C was recommended in uncomplicated surgical cases and higher doses were recommended in surgical intensive care unit patients.

In 2013, Hemilä [108] detected the significant reduction of the common cold induced asthma and respiratory infections incidence as a direct effect of oral or intravenous vitamin C in susceptible individuals.

7.4 Vitamin C dosage in old patients

In old aged individuals, vitamin C deficiency is a common feature. It is mainly attributed to improper intake within diet which is usually lower than the average recommended dietary allowance (RDA) by 25.9–38.3% [110, 111].

7.5 Vitamin C dosage in pregnant patients

In females, the hormonal changes result in increased oxidative stress. Lower vitamin C levels are usually detected in pregnant women due to several factors; obesity and iron intake which could result in low birth weight [112–114].

Up till now, there is no evidence supporting harmful effect on using ascorbic acid injection for pregnant females. Caution should be taken when administrating injection in nursing women.

7.6 Vitamin C dosage in cancer patients

The tumor cells are more sensitive to high intravenous (cytotoxic) levels of vitamin C than the normal ones [140]. At the administration of 10 g of intravenous vitamin C, a marked elevation of the extracellular concentration (1000 μmol/L) is detected which have a toxic effect on the cancer cells due to the action of the ascorbyl radicals (free radical species) [39]. On the contrary to the cancer cells, normal cells can compensate the damage occurred by these oxidative species [141]. It was also found that these mega doses of vitamin C are essential in other diseases as diabetes, cataracts, glaucoma, macular degeneration, atherosclerosis, stroke and heart diseases [40].

7.7 Vitamin C dosage in stressful patients

In people under stressful conditions either physically or emotionally, vitamin C deficiency is a common incidence. Serum levels of stress indicators were measured in pilots and reported as following; 21.1% higher malondialdehyde (MDA), 21.7% higher superoxide dismutase (SOD), and 25.1% higher total antioxidant capacity (TAC). Higher doses of vitamin C are recommended [156]. Among animals, their bodies can produce 5 times of vitamin C when exposed to stressful conditions. Doses of 30–200 times greater than the RDA of 90 mg/day are recommended during stress [157].

8. Vitamin C route of administration

Ascorbic acid is a water soluble vitamin which facilitates its absorption from buccal mucosa, gingival tissues, stomach and small intestine [14]. The literature reported several routes of vitamin C administration. In order to explain how we can reach the optimal route of administration and dosage, accurate analysis of the treated tissue condition and its nature in healthy and different pathological conditions is highly recommended.

By all routes of administration, the plasma level of vitamin C returns to its normal range within 24 h [31].

8.1 Oral route

The oral route of administration is the most common one which can be available either in the form of tablets, powder or solution. It is an essential element in multivitamins supplements [7, 31, 47].

8.2 Parental route

The intravenous route is used in advanced cancer therapy and severe illness as a complementary treatment [144]. On the contrary to oral route, the vitamin C plasma level cannot be controlled when administrated parentally. It was detected that the vitamin C plasma level is 25 fold higher than the level recorded by the oral doses [115]. It was detected that the vitamin C plasma level is 30–70 fold which is higher than the level recorded by the oral doses [116]. The urinary excretion levels were also elevated [47].

The intravenous vitamin C administration is usually used to treat hyperpigmentation especially in patients under chronic hemodialysis [47].

8.3 Topical route and topical formulations

The oral route not actually provides a source of vitamin C to peripheral structures as skin. It was detected that the vitamin C level in skin is very low. The only route that can provide a vitamin C source for skin is the topical or local routes. It was found that the usage of local application promotes the surgical healing and better tissue reconstruction [35].

Although the advantages of the topical route, the epidermal absorption of vitamin C is still limited. The water soluble nature of ascorbic acid is the main cause behind its limited penetration [60, 117, 118]. Ascorbyl palmitate and magnesium ascorbyl phosphate (buffered forms) are esters which provide higher lipid solubility through stratum corneum and higher photostability [117]. In 2002, Joung and Yi detected the low penetration power of the water soluble L-ascorbic acid into the lipophilic stratum corneum.

Acidity and concentration of topical vitamin C control its absorption. For optimal percutaneous absorption, acidic pH < 3.5 is required which is lower than the L-ascorbic acid pH (4.2) [60, 119]. Concentration of ≤20% is associated with high absorption and tissue saturation [119].

When dealing with oral mucosal tissues, vitamin C is easily absorbed by passive diffusion through the buccal mucosa [14]. It was found that the absorption of the ascorbic acid through the buccal mucosa and small intestine is nearly equal [120].

8.4 Transdermal route

Transdermal route is a big title enrolled many different techniques as sonophoresis and nanoparticles route that enhance the absorption and penetration of the topically delivered hydrophilic drugs. It can be used with insulin, morphine, caffeine, glucose, lidocaine and ascorbic acid. The micro-vibrations produced by using the ultrasound waves are the main responsible to increase the kinetic energy of the drug and deliver it deeper through the skin layers [121, 158].

8.4.1 Sonophoresis

It is one of the transdermal routes which used to increase the absorption and penetration of the topically delivered hydrophilic drugs. It can be used with insulin, morphine, caffeine, glucose, lidocaine and ascorbic acid. The micro-vibrations produced by using the ultrasound waves are the main responsible to increase the kinetic energy of the drug and deliver it deeper through the skin layers. The local and topical applications of drugs has many benefits in targeting the drug benefits directly to the area of interest and also the avoidance of the absorption, metabolism, excretion and dose problems.

In 2003, Huh et al. [45] studied the effect of iontophoresis and topical vitamin C in melasma treatment. Vitamin C solution and iontophoresis were applied. For better introduction, a low molecular weight solution was used instead of gel or cream. Significant reduction of pigmentation was detected after application.

8.4.2 Nanoparticles

L-Ascorbic acid is unstable formula of vitamin C when exposed to air, moisture, oxygen and base. The end products of L-ascorbic acid are 2,3-diketo-L-gulonic acid, oxalic acid, L-threonic acid, L-xylonic acid and L-lyxonic acid. Many researches tried to overcome this problem either by using its salts as magnesium ascorbyl phosphate or to encapsulate it using liposome, microemulsions, lipid crystals or inorganic component as hydrated ZnO. This encapsulation not interferes with the efficiency of the drug used but it controls its release and increase its penetration power [159].

8.4.3 Injection

In 2004, Senturk et al. [122] intraperitoneal injection of vitamin C was applied. Skin and serum specimens were taken at the end of the experiment. Significant reduction in the level of serum of inflammatory cytokines, IL-6 and TNF-alpha was detected. Improvement of tissue collagen was also detected.

Approximately 4 days are the detected half-life of the topically applied vitamin C (the remaining amount in tissues) [119].

9. Dosage calculation or different routes of administration

There is strong correlation between the applied dosages and the intestinal absorption rate. Marked absorption (98%) is reported in lower doses (15–30 mg). on the other hand, vitamin C absorption could be reduced to 50% in large doses that exceeds 1000 mg [15, 136, 137]. These manifestations are dose related. It can be controlled by either, reducing the total daily dose, dividing the total dose into multiple small doses, administrating the vitamin with food to decrease its absorption or to take the buffered form of vitamin C as sodium ascorbate or calcium ascorbate.

10. Contraindications of vitamin C administration

There are no contra-indications of vitamin C administration. Diabetics, patients prone to recurrent renal calculi, those undergoing stool occult blood tests, and those on sodium restricted diets or anticoagulant therapy should not take excessive doses of ascorbic acid over an extended period of time.

Diabetics taking more than 500 mg of ascorbic acid daily may obtain false readings of their urinary glucose test. No exogenous ascorbic acid should be ingested for 48–72 h before amine dependent stool occult blood tests are conducted because possible false-negative results may occur.

11. Vitamin C side effects

11.1 Oral route side effects

Side effects of vitamin C could be only detected with large doses exceeding the ULs for each individual especially on a single intake. Most of the vitamin C drawbacks were reported during oral uptake. They include diarrhea, abdominal pain [13], renal stones [123, 124] and enamel erosion during chewing [160].

11.1.1 GIT disturbance

Because of the poor oral bioavailability of vitamin C, toxic signs and symptoms may appear with large doses as a single dose. Diarrhea can occur. To avoid diarrhea occurrence, 2 g is the maximum permissible single dose [100]. Diarrhea and abdominal pain may occur due to the excretion of large amount of un-metabolized vitamin C [161]. Such manifestations are dose related. It can be controlled by either, reducing the total daily dose, dividing the total dose into multiple small doses, administrating the vitamin with food to decrease its absorption or to take the buffered form of vitamin C as sodium ascorbate or calcium ascorbate [100]. Even the usage of encapsulated vitamin C could not protect against the gastric upset. The gastrointestinal symptoms usually disappear within 1–2 weeks [19, 125].

11.1.2 Renal stones

Vitamin C metabolism results in calcium oxalate salts. Formation of renal stones (oxalate salts) and oxaluria are resulted in overdoses of vitamin C [162]. Later on, it was detected that this oxaluria is usually due to laboratory artifact occurs in the urine collection tube (ex vivo). It was also detected that vitamin C counteracting the formation of calcium oxalate crystals because of its ability to bind to calcium found in urine. Vitamin C also has the ability to increase the solubility of the calcium oxalate due to its mild acidity. It also triggers the normal urination process and prevents urine retention. All the previous actions decrease the incidence of kidney stones formation [123].

Also, it was found that vitamin C increases the urinary execration of uric acid and decreasing the plasma level of uric acid. But others demand the uricosuric effects of vitamin C as the rapid migration of uric acid from tissues. It was detected that 1 or 2 g per day increases the urinary oxalate stones were detected. Some studies detect lowering of urine pH after vitamin C intake [123].

The renal stones incidence of accumulation occurs in an average concentration of 60 g (IV) and more than 5 g (oral) [163]. In case of renal insufficiency, 1 g/day for 3 months is enough to produce renal stones (Alkhunaizi and Chan, 1996).

11.1.3 Metabolism side effects

It accelerates the absorption of other heavy metals as lead and mercury which increases its toxicity (Wyngaarded, 1987). In patients with high iron stores, vitamin C worsens the state. It also increases the iron-induced oxidative damage (Slivaka et al., 1986).

11.1.4 Dental side effects

Enamel erosions were detected on chewing of vitamin C tablets [126]. It was found that the usage of unbuffered form of ascorbic acid could result in enamel erosion [125].

11.2 Parenteral route side effects

During injection, transited mild soreness occurs during intramuscular, subcutaneous routes. Faintness or dizziness was reported on rapid intravenous administration. The renal stones incidence of accumulation occurs in an average concentration of 60 g (IV) and more than 5 g (oral) [163]. In case of renal insufficiency, 1 g/day for 3 months is enough to produce renal stones (Alkhunaizi and Chan, 1996).

12. Vitamin C deficiency

In scurvy, absence of wound healing and failure of fractured bones to heal. This was explained by the deficiency of collagen formation due to the vitamin C deficiency. Scurvy could be produced if reduction of the body reservoir of vitamin C into its fifth. The required body reservoir and the needed dosage are determinate according to the body weight [127].

In scurvy, body weakness, legs and arms edema, nose, skin and gums hemorrhage, infections, bone and cartilage damage (osteoporosis), vasculitis and cardiomegaly [164]. Many forms of bleeding found as petechiae, subperiosteal hemorrhage, ecchymoses, purpura, bleeding gums, hemarthrosis [165].

Author details

Nermin M. Yussif
Periodontology Department, MSA University, Giza, Egypt

*Address all correspondence to: dr_nermin_yusuf@yahoo.com

IntechOpen

References

[1] Pauling L. Vitamin C and the Common Cold. New York: Avon Book Company; 1970. p. 233

[2] Kleszczewska E, Misiuk W. Determination of chelate complexes by spectrophotometry; L-ascorbic acid with cadmium (II) and zinc (II) in alkaline solution. Acta Poloniae Pharmaceutica-Drug Research. 2000;**57**(5):327-330

[3] Nishikimi M, Fukuyaman R, Minoshiman I, Shimizux N, Yag K. Cloning and chromosomal mapping of the human non-functional gene for L-gulono-y-lactone oxidase, the enzyme for L-ascorbic acid biosynthesis missing in man. The Journal of Biological Chemistry. 1994;**269**(18):13685-13688

[4] Garriguet D. The effect of supplement use on vitamin C intake. Statistics Canada. 2010;**21**(1):1-7

[5] Haytowitz D. Information from USDA's nutrient data book. Journal of Nutrition. 1995;**125**:1952-1955

[6] Chandra R, Kumar S, Singh S, Sharma K, Alam N, Verma D. Quantitative assay evaluation of vitamin 'C' from formulated tablets: Application on Rp-Hplc and Uv-spectrophotometry. International Journal of Analytical, Pharmaceutical and Biomedical Sciences. 2013;**2**(3): 19-23. ISSN: 2278-0246

[7] Elmore A. Final report of the safety assessment of L-ascorbic acid, calcium ascorbate, magnesium ascorbate, magnesium ascorbyl phosphate, sodium ascorbate, and sodium ascorbyl phosphate as used in cosmetics. International Journal of Toxicology. 2005;**24**(2):51-111

[8] Velisek J, Cejpek K. Biosynthesis of food constituents: Vitamins. Water-soluble vitamins, part 2—A review. Czech Journal of Food Science. 2007;**25**:49-64

[9] Hidiroglou M, Ivan M, Batra TR. Concentration of vitamin C in plasma and milk of dairy cattle. Annales de Zootechnie. 1995;**44**:399-402

[10] Calder PC, Albers R, Antoine JM, et al. Inflammatory disease processes and interactions with nutrition. The British Journal of Nutrition. 2009;**101**(Suppl 1):S1-S45

[11] Thurnham DI. Water-soluble vitamins (vitamin C and B vitamins, thiamin, riboflavin and niacin). In: Garrow JS, James WPT, Ralph A, editors. Human Nutrition and Dietetics. Edinburgh London: Churchill. Livingston Publishers; 2000. pp. 249-257

[12] Aguirre R, May J. Inflammation in the vascular bed: Importance of vitamin C. Pharmacology and Therapeutics Journal. 2008;**119**(1):96-103

[13] Levine M, Conry-Cantilena C, Wang Y, Welch RW, Washko PW, Dhariwal KR, et al. Vitamin C pharmacokinetics in healthy volunteers: Evidence for a recommended dietary allowance. Proceedings of the National Academy of Sciences of the United States of America. 1996;**93**(8):3704-3709

[14] Stevenson NR. Active transport of L-ascoric acid in the human ileum. Gastroenterology. 1974;**67**:952-956

[15] Tsukaguchi H, Tokui T, Mackenzie B, Berger UV, Chen XZ, Wang YX, et al. A family of mammalian Na+-dependent L-ascorbic acid transporters. Nature. 1999;**399**:70-75

[16] Li Y, Schellhorn HE. New developments and novel therapeutic perspectives for vitamin C. The Journal of Nutrition. 2007;**137**(10):2171-2184

[17] Vera JC, Rivas CI, Fischbarg J, Golde DW. Mammalian facilitative hexose transporters mediate the transport of dehydroascorbic acid. Nature. 1993;**364**(6432):7982

[18] Fukuwatari T, Shibata K. Urinary water-soluble vitamins and their metabolite contents as nutritional markers for evaluating vitamin intakes in young Japanese women. Journal of Nutritional Science and Vitaminology. 2008;**54**:223-229

[19] Duconge J, Jorge R, Miranda-Massari, Gonzalez MJ, Jackson JA, Warnock W, et al. Pharmacokinetics of vitamin C: Insights into the oral and intravenous administration of ascorbate. PRHSJ. 1979;**27**(1)

[20] Pacier C, Martirosyan DM. Vitamin C: Optimal dosages, supplementation and use in disease prevention. Functional Foods in Health and Disease. 2015;**5**(3):89-107

[21] Yavorsky M, Almaden P, King CG. The vitamin C content of human tissues. Journal of Biological Chemistry. 1934;**106**:525

[22] Kalokerinos A, Dettman I, Meakin M. Endotoxin and vitamin C part 1—Sepsis, endotoxin and vitamin C. Journal of Australasian College Nutritional and Environmental Medicine. 2005;**24**(1):17-21

[23] Diaz L, Miramontes M, Hurtado P, Allen K, Avila M, de Oca E. Ascorbic acid, ultraviolet C rays and glucose but not hyperthermia are elicitors of human β-defensin 1 mRNA in normal keratinocytes. BioMed Research International. 2015;**714580**:1-9

[24] Sorice A, Guerriero E, Capone F, Colonna G, Castello G, Costantini S. Ascorbic acid: Its role in immune system and chronic inflammation diseases. Mini-Reviews in Medicinal Chemistry. 2014;**14**:444-452

[25] Weidinger A, Kozlov A. Biological activities of reactive oxygen and nitrogen species: Oxidative stress versus signal transduction. Biomolecules. 2015;**5**:472-484

[26] Neuzil J, Thomas SR, Stocker R. Requirement for, promotion, or inhibition by alpha-tocopherol of radical-induced initiation of plasma lipoprotein lipid peroxidation. Free Radical Biology & Medicine. 1997;**22**:57-71

[27] Mansour MA, Nagi MN, El-Katip AS, Al_Bekairi AM. Effect of tymoquinone on antiioxidant enzyme activities lipid peroxidation and DT-diaphorase in different tissues of mice: A possible mechanism of action. Cell Biochemistry and Function. 2002;**20**:143-151

[28] Popovic L, Mitic N, Miric D, Bisevac B, Miric M, Popovic B. Influence of vitamin C supplementation on oxidative stress and neutrophil inflammatory response in acute and regular exercise. Oxidative Medicine and Cellular Longevity. 2015;**295497**:1-7

[29] Sharma A, Sharma S. Reactive oxygen species and antioxidants in periodontics: A review. International Journal of Dental Clinics. 2011;**3**(2):44-47

[30] Valko M, Rhodes CJ, Moncol J, Izakovic M, Mazur M. Free radicals, metals and antioxidants in oxidative stress-induced cancer. Chemico-Biological Interactions. 2006;**160**(1):1-40

[31] Padayatty S, Katz A, Wang Y, Eck P, Kwon O, Lee J, et al. Vitamin C as an antioxidant: Evaluation of its role in disease prevention. Journal of the American College of Nutrition. 2003;**22**(1):18-35

[32] Devlin TM. Textbook of Biochemistry with Clinical Correlation.

5th ed. New York: Wiley-Liss, A John Wiley and Sons, Inc. Publication; 2002. pp. 121-126, 590-593, 1157-1159

[33] May J. Is ascorbic acid an antioxidant for the plasma membrane? The Journal of Federation of American Societies for Experimental Biology. 1999;**13**:995-1006

[34] Prockop D, Kivirikko K. Collagens: Molecular biology, diseases, and potentials for therapy. Annual Review of Biochemistry. 1995;**64**:403-434

[35] Farris P. Cosmeceutical vitamins: Vitamin C. In: Procedures in Cosmetic Dermatology Series: Cosmeceuticals. Philadelphia: Elsevier; 2005. pp. 51-56

[36] Alster T, West T. Effect of topical vitamin c on postoperative carbon dioxide laser resurfacing erythema. Dermatologic Surgery. 1998;**24**(3):331-334

[37] Halliwell B. Biologically significant scavenging of the myeloperoxidase-derived oxidant hypochlorous acid by ascorbic acid: Implications for antioxidant protection in the inflamed rheumatoid joint. FEBS Letters. 1987;**213**:15-17

[38] Davis RH. Vitamin C influence on localized adjuvant arthritis. Journal of the American Podiatric Medical Association. 1990;**80**:414

[39] Sakagami H, Satoh K, Hakeda Y, Kumegawa M. Apoptosis-inducing activity of vitamin C and vitamin K. Cellular and Molecular Biology (Noisy-le-Grand, France). 2000;**46**:129-143

[40] Will JC, Tyers T. Does diabetes mellitus increase the requirement for vitamin C? Nutrition Reviews. 1996;**54**:193-202

[41] Tewary A, Patra BC. Use of vitamin C as an immunostimulant.

Effect on growth, nutritional quality, and immune response of *Labeo rohita* (Ham.). Fish Physiology and Biochemistry. 2008;**34**:251-259

[42] Cadenas S, Cadenas AM. Fighting the strangerantioxidant protection against endotoxin toxicity. Toxicology. 2002;**180**(1):45-63

[43] Parvez S, Kang M, Chung H, Cho C, Hong M, Shin M, et al. Survey and mechanism of skin depigmenting and lightening agents. Phytotherapy Research. 2006;**20**:921-934

[44] Choi YK, Rho YK, Yoo KH, Lim YY, Li K, Kim BJ, et al. Effects of vitamin C vs. multivitamin on melanogenesis: Comparative study in vitro and in vivo. International Journal of Dermatology. 2010;**49**(2):218-226. DOI: 10.1111/j.1365-4632.2009.04336.x

[45] Huh C, Seo K, Park J, Lim J, Eun H, Park K. A randomized, double-blind, placebo-controlled trial of vitamin C iontophoresis in melasma. Dermatology. 2003;**206**(4):316-320

[46] Lee G. Intravenous vitamin C in the treatment of post-laser hyperpigmentation for melasma: A short report. Journal of Cosmetic and Laser Therapy. 2008;**10**:234-236

[47] Padayatty S, Sun H, Wang Y, Riordan H, Hewitt S, Katz A, et al. Vitamin C pharmacokinetics: Implications for oral and intravenous use. Annals Internal Medicine. 2004;**140**:533-537

[48] Panich U, Tangsupa-a-nan V, Onkoksoong T, Kongtaphan K, Kasetsinsombat K, Akarasereenont P, et al. Inhibition of UVA-mediated melanogenesis by ascorbic acid through modulation of antioxidant defense and nitric oxide system. Archives of Pharmcoceutical Researches. 2011;**34**(5):811-820

[49] Kameyama KC et al. Inhibitory effect of magnesium-L-ascorbyl-2-phosphate (VC-PMG) on melanogenesis in vitro and in vivo. Journal of the American Academy of Dermatology. 1996;**34**:29-33

[50] Morganti P, Fabrizi G, James B. An innovative cosmeceutical with skin whitening activity. Journal of Applied Cosmetology. 1999;**17**(4):144-153

[51] Park K, Huh S, Choi1 H, Kim D. Biology of melanogenesis and the search for hypopigmenting agents. Dermatologica Sinica. 2010;**28**:53-58

[52] Lloyd H, Kammer J. Treatment of hyperpigmentation. Seminars in Cutaneous Medicine and Surgery. 2011;**30**:171-175

[53] Solano F, Briganti S, Picardo M, Ghanem G. Hypopigmenting agents: An updated review on biological, chemical and clinical aspects. Pigment Cell Research. 2006;**19**:550-571

[54] Candelario-Jalil E, Akundi RS, Bhatia HS, Lieb K, Appel K, Muñoz E, et al. Ascorbic acid enhances the inhibitory effect of aspirin on neuronal cyclooxygenase-2-mediated prostaglandin E2 production oral contraceptive pills. Journal of Neuroimmunology. 2006;**174**(1-2):39-51

[55] Wilson M, Baguley B, Wall C, Jameson M, Findlay M. Review of high-dose intravenous vitamin C as an anticancer agent. Asia-Pacific Journal of Clinical Oncology. 2014;**10**:22-37

[56] Kang S, Cho D, Kim Y, Hahm E, Kim Y, Jin S, et al. Sodium ascorbate (vitamin C) induces apoptosis in melanoma cells via the down-regulation of transferrin receptor dependent iron uptake. Journal of Cellular Physiology. 2005;**204**:192-197

[57] Stephenson C, Levin R, Spector T, Christopher G. LisPhase I clinical trial to evaluate the safety, tolerability, and pharmacokinetics of high-dose intravenous ascorbic acid in patients with advanced cancer. Cancer Chemotherapy and Pharmacology. 2013;**72**:139-146

[58] Choy CKM, Benzie IFF, Cho P. Antioxidants in tears and plasma: Inter-relationships and effect of vitamin C supplementation. Current Eye Research. 2003;**27**:55-60

[59] Asarian J, Zimmerman M. Antiaging secrets from a to z. 2001. www.zgrouponline.com/ebook/antiaging.pdf

[60] Fitzpatrick R, Rostan E. Double blind, half face study comparing topical vitamin C and vehicle for rejuvenation of photodamage. Dermatological Surgical. 2002;**28**(3):231-236

[61] Naidu KA. Vitamin C in human health and disease is still a mystery? An overview. Nutrition Journal. 2003;**2**:7

[62] Zanoni J, Lucas N, Trevizan A, Souza I. Histological evaluation of the periodontal ligament from aged Wistar rats supplemented with ascorbic acid. Anais de Academia Brasileina de Ciencias. 2015;**85**(1):327-335

[63] Kim Y, Kim K, Lim S, Choi S, Moon J, Kim J, et al. Favorable effect of dietary vitamin C on bone mineral density in postmenopausal women (KNHANES IV, 2009): Discrepancies regarding skeletal sites, age, and vitamin D status. Osteoporosis International. 2015;**26**(9):2329-2337

[64] Nusgens B, Humbert P, Rougier A, Colige A, Haftek M, Lambert C, et al. Are topically applied vitamin C enhances the mRNA level of collagens I and III, their processing enzymes and tissue inhibitor of matrix metalloproteinase 1 in the human dermis. The Journal of Investigative Dermatology. 2001;**116**(6):854-859

[65] Zhang J-Y, Doll BA, Beckman EJ, Hollinger JO. Tissue Engineering. 2003;9(6):1143-1157

[66] Urban K, Holing HJ, Lüttenberg B, Szuwart T, Plate U. An in vitro study of osteoblast vitality influenced by the vitamins C and E. Head & Face Medicine. 2012;8:25

[67] Stojiljkovic D, Pavlovic D, Arsic I. Oxidative stress, skin aging and antioxidant therapy. Scientific Journal of the Faculty of Medicine in Nis. 2014;31(4):207-217

[68] Sachs D, Ritti L, Chubb H, Orringer J, Fisher G, Voorhees J. Hypo-collagenesis in photoaged skin predicts response to anti-aging cosmeceuticals. Journal of Cosmetic Dermatology. 2013;12:108-115

[69] Berger M, Mette M, Straaten O, Heleen M. Vitamin C supplementation in the critically ill patient. Current Opinion of Clinical Nutrition and Metabolic Care. 2015;18(2):193-201

[70] Shiga M, Kapila Y, Zhang Q, Yami T, Kapila S. Ascorbic acid induces collagenase-1 in human periodontal ligament cells but not in mct3-e1 osteoblast-like cells: Potential association between collagenase expression and changes in alkaline phosphatase phenotype. Journal of Bone and Mineral Research. 2003;18(1):67-77

[71] Tiedtke J, Marks DO, Morel J. Cosmetochem international, Switzerland stimulation of collagen production in human fibroblasts. Cosmetic Science Technology. 2007

[72] Lynch SR. Interaction of iron with other nutrients. Nutrition Reviews. 1997;55:102-110

[73] Rendón-Ramírez AL, Maldonado-Vega M, Quintanar-Escorza MA, Hernández G, Arévalo-Rivas BI, Zentella-Dehesa A, et al. Effect of vitamin E and C supplementation on oxidative damage and total antioxidant capacity in lead-exposed workers. Environmental Toxicology and Pharmacology. 2014;37(1):45-54 (cited in Pacier 2015)

[74] Zeraati F, Araghchian M, Farjoo M. Ascorbic acid interaction with analgesic effect of morphine and tramadol in mice. Anesthesiology and Pain Medicine. 2014;4(3):e19529

[75] Al-Katib S, Al-Kaabi M, Al-Jashamy K. Effects of vitamin C on the endometrial thickness and ovarian hormones of progesterone and estrogen in married and unmarried women. American Journal of Research Communication. 2013;1(8):24-23

[76] Pararajasegaram G, Sevanian A, Rao NA. Suppression of S antigen-induced uveitis by vitamin E supplementation. Ophthalmic Research. 1991;23:121-127

[77] Boobis AR, Burley D, Davies DM, et al. Therapeutic Drugs. London: Churchill Livingstone; 1991

[78] Cobley J, McHardy H, Morton J, Nikolaidis M, Close G. Influence of vitamin C and vitamin E on redox signaling: Implications for exercise adaptations. Free Radical Biology & Medicine. 2015;2(84):65-76. DOI: 10.1016/j.freeradbiomed.2015.03.018

[79] Jain H, Mulay S. A review on biological functions and sources of anti-scorbutic factor: Vitamin C. Donnish Journal of Medicinal Plant Research. 2014;1(1):001-008

[80] Niki E, Noguchi N, Tsuchihashi H, Gotoh N. Interaction among vitamin C, vitamin E and B-carotene. The American Journal of Clinical Nutrition. 1995;62:1322s-1326s

[81] Kalesh DG, Mallikarjuneswara VR, Clemetson CAB. Effect of

estrogen containing contraceptives on platelet and plasma ascorbic acid concentrations. Contraception. 1971;**4**:183-192

[82] Cathcart RIII. The method of determining proper doses of vitamin C for the treatment of disease by titrating to bowel tolerance orthomolecular. Psychiatry. 1981;**10**(2):125-132

[83] Ancha HR, Kurella RR, McKimmey CC, Lightfoot S, Harty RF. Luminal antioxidants enhance the effects of mesalamine in the treatment of chemically induced colitis in rats. Experimental Biology and Medicine. 2008;**233**:1301-1308

[84] Morrow JD. The isoprostanes: Their quantification as an index of oxidant stress status in vivo. Drug Metabolism Reviews. 2000;**32**:377-385

[85] Stampfer MJ, Hennekens CH, Manson JE, et al. Vitamin E consumption and the risk of coronary disease in women. The New England Journal of Medicine. 1993;**328**:1444-1449

[86] Belo M, de Moraes J, Soares V, Martins M, Brum C, de Moraes F. Vitamin C and endogenous cortisol in foreign-body inflammatory response in pacus. Pesquisa Agropecuaria Brasileira, Brasilia. 2012;**47**(7):1015-1021

[87] Hasselholt S, Tveden-Nyborg P, Lykkesfeldt J. Distribution of vitamin C is tissue specific with early saturation of the brain and adrenal glands following differential oral dose regimens in Guinea pigs. British Journal of Nutrition. 2015;**13**:1-11

[88] Padayatty S, Doppman J, Chang R, Wang Y, Gill J, Papanicolaou D, et al. Human adrenal glands secrete vitamin C in response to adrenocorticotrophic hormone. American Journal of Clinical Nutrition. 2007;**6**:145-149

[89] Ipsen D, Tveden-Nyborg P, Lykkesfeldt J. Does vitamin C deficiency promote fatty liver disease development? Nutrients. 2014;**6**:5473-5499

[90] Achuba F. Effect of Vitamins C and E intake on blood lipid concentration, lipid peroxidation, superoxide dismutase and catalase activities in rabbit fed petroleum contaminated diet. Pakistan Journal of Nutrition. 2005;**4**(5):330-335

[91] Christie-David D, Girgis C, Gunton J. Effects of vitamins C and D in type 2 diabetes mellitus. Nutrition and Dietary Supplements. 2015;**7**:21-28

[92] Yilmaz C, Erdemli E, Selek H, Kinik H, Arikan M, Erdemli B. The contribution of vitamin C to healing of experimental fractures. Archives of Orthopaedic and Trauma Surgery. 2001;**121**:426-428

[93] Ganta DR, McCarthy MB, Gronowicz GA. Ascorbic acid alters collagen integrins in bone culture. Endocrinology. 1997;**138**:3606-3612

[94] Kipp DE, Grey CE, McElvain ME, et al. Long-term low ascorbic acid intake reduces bone mass in Guinea pigs. The Journal of Nutrition. 1996a;**126**:2044-2049

[95] Kipp DE, McElvain M, Kimmel DB, et al. Scurvy results in decreased collagen synthesis and bone density in the Guinea pig animal model. Bone. 1996b;**18**:281

[96] Weinstein M, Babyn P, Zlotkin S. An orange a day keeps the doctor away: Scurvy in the year 2000. Pediatrics. 2001;**108**:E55

[97] Otsuka E, Kato Y, Hirose S, Hagiwara H. Role of ascorbic acid in the osteoclast formation: Induction of osteoclast differentiation factor with formation of the extracellular

collagen matrix. Endocrinology. 2000;**141**:3006-3011

[98] Witenberg B, Kalir HH, Raviv Z, Kletter Y, Kravtsov V, Fabian I. Inhibition by ascorbic acid of apoptosis induced by oxidative stress in HL-60 myeloid leukemia cells. Biochemical Pharmacology. 1999;**57**:823-832

[99] Berger M. Vitamin C requirements in parenteral nutrition. Gastroenterology. 2009;**137**:70-78

[100] National Academy of Sciences. Institute of Medicine, Food and Nutrition Board. Dietary Reference Intakes for Vitamin C, Vitamin E, Selenium, and Carotenoids. Washington, D.C.: National Academy Press; 2000. pp. 1-20

[101] Levine M, Rumsey SC, Daruwala R, et al. Criteria and recommendations for vitamin C intake. JAMA. 1999;**281**:1415-1423

[102] Johnston CS, Luo B. Comparison of the absorption and excretion of three commercially available sources of vitamin C. Journal of the American Dietetic Association. 1994;**94**:779

[103] Blanchard J, Tozer TN, Rowland M. Pharmacokinetic perspectives on megadoses of ascorbic acid [see comments]. The American Journal of Clinical Nutrition. 1997;**66**(5):116571

[104] Estanol M, Crisp C, Oakley S, Kleeman S, Fellner A, Pauls R. Systemic markers of collagen metabolism and vitamin C in smokers and non-smokers with pelvic organ prolapsed. European Journal of Obstetrics and Gynecology and Reproductive biology. 2015;**184**:58-64

[105] Sargeant L, Jaeckel A, Wareham N. Interaction of vitamin C with the relation between smoking and obstructive airways disease in EPIC

Norfolk. The European Respiratory Journal. 2000;**16**:397-402

[106] Alberg AJ. The influence of cigarette smoking on circulating concentrations of antioxidant micronutrients. Toxicology. 2002;**180**(2):121-137

[107] Eiserich JP, tan der Vliet A, Handelman GJ, Halliwell B, Cross CE. Dietary antioxidants and cigarette smoke-induced biomolecular damage: A complex interaction. The American Journal of Clinical Nutrition. 1995;**62**(6):1490S-1500S

[108] Hemilä. Vitamin C and common cold-induced asthma: A systematic review and statistical analysis. Allergy, Asthma & Clinical Immunology. 2013;**9**:46

[109] Tanaka H, Matsuda T, Miyagantani Y, et al. Reduction of resuscitation fluid volumes in severely burned patients using ascorbic acid administration: A randomized, prospective study. Archives of Surgery. 2000;**135**:326-331

[110] Hsiao PY, Mitchell DC, Coffman DL, Allman RM, Locher JL, Sawyer P, et al. Dietary patterns and diet quality among diverse older adults: The University of Alabama at Birmingham study of aging. The Journal of Nutrition, Health & Aging. 2013;**17**(1):19-25

[111] Sharkey JR, Branch LG, Zohoori N, Giuliani C, Busby-Whitehead J, Haines PS. Inadequate nutrient intakes among homebound elderly and their correlation with individual characteristics and health-related factors. The American Journal of Clinical Nutrition. 2002;**76**(6):1435-1445

[112] Anetor JI, Ajose OA, Adeleke FN, Olaniyan-Taylor GO, Fasola FA. Depressed antioxidant status in pregnant women on iron supplements: Pathologic and clinical correlates. Biological Trace Element Research. 2010;**136**(2):157-170

[113] Saker M, Soulimane Mokhtari N, Merzouk SA, Merzouk H, Belarbi B, Narce M. Oxidant and antioxidant status in mothers and their newborns according to birthweight. European Journal of Obstetrics, Gynecology, and Reproductive Biology. 2008;**141**(2):95-99

[114] Sen S, Iyer C, Meydani SN. Obesity during pregnancy alters maternal oxidant balance and micronutrient status. Journal of Perinatology. 2014;**34**:105-111

[115] Padayatty SJ, Riordan HD, Hewitt SM, Katz A, Hoffer LJ, Levine M. Intravenously administered vitamin C as cancer therapy: Three cases. CMAJ. 2006;**174**(7):937-942

[116] Padayatty SJ, Levine M. Vitamin C and coronary microcirculation. Circulation. 2001;**103**:E117

[117] Austria R, Semenzato A, Bettero A. Stability of vitamin C derivatives in solution and topical formulations. Journal of Pharmaceutical and Biomedical Analysis. 1997;**15**(6):795-801

[118] Colven R, Pinnell S. Topical vitamin C in aging. Clinics in Dermatology. 1996;**14**:27-234

[119] Pinnell SR et al. Topical ʟ-ascorbic acid: Percutaneous absorption studies. Dermatologic Surgery. 2001;**27**(82):137-142

[120] Sadoogh-Abasian F, Evered DF. Absorption of vitamin C from the human buccal cavity. British Journal of Nutrition. 1979;**42**(1):15-20

[121] Polat B, Blankschtein D, Langer R. Low-frequency sonophoresis: Application to the transdermal delivery of macromolecules and hydrophilic drugs. Expert Opinion on Drug Delivery. 2010;7(12):1415-1432. DOI: 10.1517/17425247.2010.538679

[122] Senturk N, Keles GC, Kaymaz FF, Yildiz L, Acikgoz G, Turanli AY. The role of ascorbic acid on collagen structure and levels of serum interleukin-6 and tumour necrosis factor-alpha in experimental lathyrism. Clinical and Experimental Dermatology. 2004;**29**(2):168-175

[123] Baxmann A, Mendonc C, Heilberg I. Effect of vitamin C supplements on urinary oxalate and pH in calcium stone-forming patients. Kidney International Journal. 2003;**63**:1066-1071

[124] Dinca A, Shova S, Ion A, Maxim C, Lloret F, Julve M, et al. Ascorbic acid decomposition into oxalate ions: A simple synthetic route towards oxalato-bridged heterometallic 3d-4f clusters. Dalton Transactions. 2015;**44**(16):7148-7151

[125] Fleming D, Tucker KL, Jacques PF, Dallal GE, Wilson PWF, Wood RJ. Dietary factors associated with the risk of high iron stores in the elderly Framingham heart study cohort. The American Journal of Clinical Nutrition. 2002;**76**:1375-1384

[126] Giunta J. Dental erosion resulting from chewable vitamin C tablets. Journal of the American Dental Association (1939). 1983;**07**(2):253-256

[127] Rathee M. Vitamin C and oral health: A review. Indian Journal of Applied Research. 2013;**3**(9):1-2

[128] Mehlhorn RJ. Ascorbate- and dehydroascorbic acid-mediated reduction of free radicals in the human erythrocyte. Journal of Biological Chemistry. 1991;**266**:2724-2731

[129] Galesso M, Gatta M, Galiano F. Comparative studies on the stability of ascorbic acid and its derivatives in various matrixes and interaction with commonly used cosmetic preservatives. Cosmet Toiletries Journal. 1993;**2**:58-74

[130] Laki K. Das ox-redoxpotential der ascorbinsäure. Zeitschrift für Physikalische Chemie. 1933;**217**:54

[131] Green DE. The potentials of ascorbic acid. Biochemical Journal. 1933;**27**(1044):7

[132] Olmsted JMD. American Journal of Physiology. 1935;**3**:551

[133] Zilva SS. Biochemical Journal. 1928;**22**:779

[134] Demole I. Zeitschrift fur Physiologische Chemie. 1933;**217**:83

[135] Duarte T, Lunec J. Review: When is an antioxidant not an antioxidant? A review of novel actions and reactions of vitamin C. Free Radical Research. 2005;**39**(7):671-686. DOI: 10.1080/10715760500104025

[136] Ginter E. Vitamin C deficiency, cholesterol metabolism and atherosclerosis. Journal of Orthomolecular Medicine. 1982;**6**(3 & 4):1991

[137] Brennan LA, Morris GM, Wasson GR, Hannigan BM, Barnett YA. The effect of vitamin C or vitamin E supplementation on basal and H_2O_2-induced DNA damage in human lymphocytes. British Journal of Nutrition. 2000;**84**:195-202

[138] Tsao CS, Miyashita K. Effects of high intake of ascorbic acid on plasma levels of amino acids. IRCS Journal of Medical Science. 1984;**12**:1052-1053

[139] Frank E, Bendich A, Denniston M. Use of vitamin-mineral supplements by female physicians in the United States. American Journal of Clinical Nutrition. 2000;**72**:969-975

[140] Maramag C, Menon M, Balaji KC, Reddy PG, Laxmanan S. Effects of vitamin C on prostate cancer cells in vitro: effects on cell number,

viability and DNA synthesis. Prostate. 1997;**32**(3):188-195

[141] Grad JM, Bahlis NJ, Reis I, Oshiro MM, Dalton WS, Boise LH. Ascorbic acid enhances arsenic trioxide-induced cytotoxicity in multiple myeloma cells. Blood. 2001;**98**:805-813

[142] Gaby S, Singh V. Vitamin C. In: Gaby SK, Bendich A, Singh V, Machlin L, editors. Vitamin Intake and Health: A Scientific Review. NY: Marcel Dedder; 1991

[143] Voigt K, Kontush A, Stuerenburg HJ, Muench-Harrach D, Hansen HC, Kunze K. Decreased plasma and cerebrospinal fluid ascorbate levels in patients with septic encephalopathy. Free Radical Research. 2002;**36**:735-739

[144] Riordan NH, Riordan HD, Casciari JJ. Clinical and experimental experiences with intravenous vitamin C. Journal of Orthomolecular Medicine. 2000;**15**:201-213

[145] Pérez-Vicente A, Serrano P, Abellán P, García-Viguera C. Influence of packaging material on pomegranate juice colour and bioactive compounds, during storage. Journal of the Science of Food and Agriculture. 2004;**84**(7):639-644

[146] Cameron E, Pauling L. Supplemental ascorbate in the supportive treatment of cancer: Prolongation of survival times in terminal human cancer. Proceedings of the National Academy of Sciences of the United States of America. 1976;**73**:3685-3689

[147] Weber C. Increased adhesiveness of isolated monocytes to endothelium is prevented by vitamin c intake in smokers. Circulation. 1996;**93**:1488-1492

[148] Mehta JB, Singhal SB, Mehta BC. Ascorbic-acid-induced haemolysis

in G-6-PD deficiency. Lancet. 1990;**336**:944

[149] Sauberlich HE. Pharmacology of vitamin C. Annual Review of Nutrition. 1994;**14**:371-391

[150] Fu S, Wai-Hang Cheng A, Yau-Chuk Cheuk A, Tsui-Yu Mok A, Christer Rolf B, Shu-Hang Yung A, et al. Development of vitamin C irrigation saline to promote graft healing in anterior cruciate ligament reconstruction. Journal of Orthopaedic Translation. 2013;**1**:67e77

[151] Belo MAA, Schalch SHC, Moraes FR, Soares VE, Otoboni A, Moraes JER. Effect of dietary supplementation with vitamin E and stocking density on macrophage recruitment and giant cell formation in the teleost fish, *Piaractus mesopotamicus*. Journal of Comparative Pathology. 2005;**133**:146-154

[152] Tort L. Stress and immune modulation in fish. Developmental and Comparative Immunology. 2011;**35**:1366-1375

[153] Ministry of Health, Labor and Welfare: dietary reference intakes for Japanese. 2010

[154] Long CL, Maull KI, Krishnan RS, et al. Ascorbic acid dynamics in the seriously ill and injured. Journal of Surgical Research. 2003;**109**:144-148

[155] Fukushima R, Koide T, Yamazaki E, et al. Water-soluble-vitamin status in the postoperative gastrointestinal surgical patients receiving peripheral parenteral nutrition. Clinical Nutrition. 2009;**4**:164

[156] Taleghani EA, Sotoudeh G, Amini K, Araghi MH, Mohammadi B, Yeganeh HS. Comparison of antioxidant status between pilots and non-flight staff of the Army Force: Pilots may need more vitamin C. Biomedical and Environmental Sciences. 2014;**27**(5):371-377

[157] Chatterjee IB, Majunder AK, Nandi BK, Subramadian N. Synthesis and some major functions of vitamin C in animals. Annals of the New York Academy of Sciences. 1975;**258**:24-47

[158] Ngamratanapaiboon S, Iemsan-Arng J, Yambangyang P, Neatpisarnvanit C, Sirisoonthorn S, Sathirakul K. In vitro study the transdermal permeation profiles of L-ascorbic acid in chitosan hydrogel formulation altered by sonophoresis. Advance Journal of Pharmaceutical Sciences. 2012;**1**(1):13-17

[159] Yamamoto N, Kabuto H, Matsumoto S, Ogawa N, Yokoi I. α-Tocopheryl-L-ascorbate-2-O-phosphate diester, a hydroxyl radical scavenger, prevents the occurrence of epileptic foci in a rat model of post-traumatic epilepsy. Pathophysiology. 2002;**8**:205-214

[160] Pandey N, Rai S. Biochemical Activity and Therapeutic Role of Antioxidants in Plants and Humans. 2014

[161] Wandzilak TR, D'Andre SD, Davis PA, Williams HE. Effect of high dose vitamin C on urinary oxalate levels. Journal of Urology. 1994;**151**:834-837

[162] Urivetsky M, Kessaris D, Smith AD. Ascorbic acid overdosing: A risk for calcium oxalate nephrolithiasis. Journal of Urology. 1992;**147**:1215-1218

[163] Mashour S, Turner JF Jr, Merrell R. Acute renal failure, oxalosis, and vitamin C supplementation: A case report and review of the literature. Chest. 2000;**118**:561-563

[164] Packer L, Fuchs J. Vitamin C in Health and Disease. New York, Basel, Hong Kong: Marcel Dekker, Inc; 1997

[165] Anonymous. Ascorbic acid intake and salivary ascorbate levels. Nutrition Reviews. 1986;**44**:328-330

Chapter 3

Vitamin C: An Epigenetic Regulator

Fadime Eryılmaz Pehlivan

Abstract

Vitamin C is an essential micronutrient, a free radical scavenger; while it has functions such as blocking oncogenic transformation induced by carcinogens. A different view of the potential action of vitamin C in cancer came from the discovery of its importance for activation of ten-eleven translocation (TET) enzymes that are involved in demethylation of DNA and histones. Aberrant DNA and histone methylation are hallmarks of all cancers and may result from altered expression or point mutations in the genes encoding these regulatory enzymes. Recent studies have shown that vitamin C potentiates the effects of DNA methyltransferase inhibitors. Epigenetic alterations, along with genetic mutations, are known to contribute to onset of cancer. Vitamin C is found to be a key mediator of the interface between genome and environment, regulating DNA demethylation as a cofactor for TET dioxygenases. It is shown that vitamin C drives active removal of DNA methylation by enhancing TET enzymes, which helps to erase DNA methylation and epigenetic memory encoded by it to improve reprogramming of differentiated cells to an embryonic-like state. Here, an overview of the role of vitamin C as an essential factor for epigenetic regulation and its potential in epigenetic therapy in cancer patients is provided.

Keywords: vitamin C, epigenesis, DNA modulation, TET enzymes

1. Vitamin C

1.1 Vitamin C, free radicals and antioxidant mechanism

Vitamin C (L-ascorbic acid) is a multifunctional water-soluble antioxidant substance in both plants and animals, having vital and ubiquitous roles in the life processes. It was first isolated and characterized by Szent-Gyorgyi back in 1928 [1]. In plants especially in green leaves, this metabolite is one of the most abundant, representing 10% of the total soluble carbohydrate pool [2]. In both plants and animals, it has important roles as a cofactor for a large number of key enzymes, influencing mitosis and cell growth by modulating the expression of specific genes involved in defense and hormonal signaling pathways. It is also required for iron utilization, connective tissue, cardiovascular functions, and immune cell development [3]. The disease, scurvy, which is defined as a medical situation caused by lack of L-ascorbic acid has considerable historical significance in the discovery of this amazing substance. In nineteenth century sailors and others who did not have access to natural sources of vitamin C (fresh vegetables and fruits) suffered from bleeding under the skin, severe ulcers, depression, loss of teeth and joint weakness [1].

In an effort to find a cure, scientists discovered that by consuming certain vegetables and fruits, this condition can be treated and prevented from reoccurring [1]. Plants and most animals synthesize vitamin C on their own, but humans lack this ability due to a deficiency in an enzyme called "L-gulono-gamma-lactone oxidase" [1–3] that catalyzes the final step in vitamin C biosynthesis, representing that humans must obtain this vital compound from exogenous sources such as natural diet and supplements.

Electrons prefer to pair up with each other so that they can remain energetically stable. However, due to biotic and abiotic influences such as infection, wounding, UV radiation, air pollutants, smoking, and natural oxygen metabolism (aerobic respiration), this balance is disrupted. Consequently various molecules can end up with a single electron in their outermost orbital, making them highly unstable and reactive towards their surroundings [4]. The newly formed molecule is now desperately looking for another electron to balance and stabilize itself with. This is what is known as a "free radical," that are urgently searching for electrons and stealing them from other molecules. When a molecule loses an electron to a free radical molecule, it becomes a free radical itself and should now find an electron from another molecule to energetically stabilize itself [4, 5]. By this way, a single free radical can initiate a chain reaction that can result in severe damage to cell membranes, organelles, and structural and genetic coding (DNA and RNA) of body [4, 5]. These destructive mechanisms result in chronic free radical assault (oxidative stress). On the contrary, living organisms have a large arsenal of active substances such as antioxidants, to scavenge them and fight back [4, 5]. Vitamin C is a highly potent (reducing agent) antioxidant capable of neutralizing reactive oxygen species (ROS) and nitrogen derived free radical species [4, 5], achieving this by donating an electron to free radicals (**Figure 1**). As being a small, water soluble, reductone sugar acid with antioxidant properties, vitamin C acts as a primary substrate for detoxification of a number of ROS such as H_2O_2, and neutralize it to superoxide radicals (O_2^{-}), singlet oxygen (O^{-}) or hydroxyl radical (OH^{-}) by acting as a secondary antioxidant during reductive recycling of the oxidized form of α-tocopherol [2, 4] (**Figure 1**).

1.2 Vitamin C and cancer

The carcinogenic effect of oxidative stress is focused on the genotoxicity of ROS. They can cause cancer through multiple mechanisms, and are known to play significant roles in the promotional stage of carcinogenesis [2, 4, 5]. Vitamin C is associated with its protective roles against ROS dependent oxidative stress, by stimulating immune function, inhibiting nitrosamine formation, and blocking the metabolic activation of carcinogens. It can also protect against oxidative DNA damage, that is implicated in tumor initiation [5, 6]. Although oxidative stress causes oxidative damage, moderate amounts of ROS are found to serve as a secondary messenger in the intracellular signaling cascades [5, 6]. Low oxidative stress is proved to be essential for cellular signal transduction that leads to the induction of detoxification or antioxidant enzyme systems. Limited amounts of ROS are needed for triggering the antioxidant signal transduction [5, 6]. Some phytochemicals are proved to induce phase II detoxification of antioxidant enzymes by triggering nuclear translocation of the transcription factors such as NF-E$_2$–related factor 2 (Nrf2) binding to antioxidant response element [7]. Many of the inducers are capable of activating these transcription factors mimic prooxidants; interestingly, most of them are antioxidants by nature [8]. Besides the antioxidant activity, prooxidant potential of vitamin C is contributed to its chemopreventive properties [8, 9]. Individuals

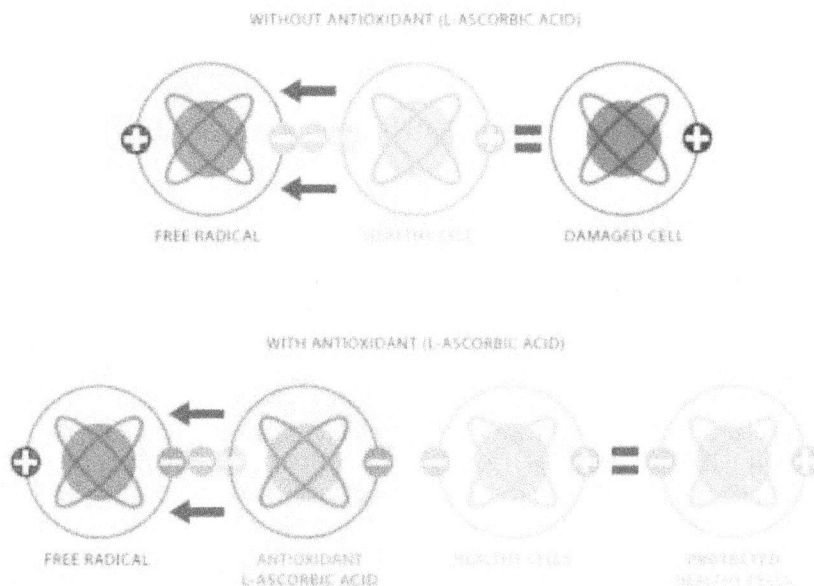

Figure 1.
Free radicals and antioxidant mechanism [2].

with cancer are reported to have significantly lower plasma vitamin C levels than healthy individuals [9]. Cancer-induced ROS formation and oxidative stress is one explanation for such a deficiency [10]. In this case, the prooxidant activity of vitamin C needs to be clarified while it is fighting, as an antioxidant, against oxidative stress. As a consequence, it is worthwhile to examine if vitamin C can induce the expression of phase II detoxification via its prooxidant potential, indicating that high intakes of vitamin C concentrations in serum are inversely associated with the risk of some cancers [10, 11].

Vitamin C at high doses has a very different (pro-oxidative) effect from vitamin C at physiological doses, that at high doses and under certain conditions (in the presence of redox-active metal ions) vitamin C can give rise to ROS, protein glycation, and DNA damage [11–13]. Increased levels of ROS and redox-active iron led to alterations in cancer cell oxidative metabolism resulting in selective sensitivity and pro-oxidative toxicity of high-dose vitamin C [11, 12]. It is reported that exposure to high-dose vitamin C sensitized cancer cells to ionizing radiation combined with chemotherapy [10–13].

1.3 Vitamin C and anti-inflammatory activity

It is evident that ROS are involved in chronic inflammation and cancer [14]. The generation of oxidative stress is an important part of the inflammatory response that is associated with tumor promotion. It is indicated that stomach cancer is a consequence of chronic inflammation [15]. This inflammatory process caused by the overproduction of ROS could be a target of vitamin C [15]. Vitamin C was proved to reduce inflammation caused by ROS [15], a recent study showed low vitamin C concentrations in gastric juice in the earlier stage of carcinogenesis [16]. Vitamin C protects gastric carcinogenesis by scavenging of the mucosal free radicals [15, 16] and by inhibiting the formation of carcinogenic nitrosamines [15, 16]. It is reported that, treatment of endothelial cells with vitamin C resulted in the accumulation of a large amount of this substance inside the cells, which consequently decreased both the intracellular free radical status and inducible nitric oxide synthase induction [17]. During the inflammation process, vitamin C is shown to inactivate nuclear factor-B in endothelial cells besides of its antioxidant activity [18], indicating the anti-inflammatory activity of vitamin C with its intrinsic antioxidant activity [16–18].

1.4 Vitamin C and cell-to-cell communication

In multicellular organisms, cell-to-cell communication through gap junction channels is essential for maintaining homeostatic balance through modulation of cell proliferation and differentiation [19]. Inhibition of this communication is related to the carcinogenic process, especially to tumor promotion [20]. A well-known tumor promoter, hydrogen peroxide, is also found to inhibit gap junction intercellular communication (GJIC) [21]. It is found that the inhibition of GJIC is strongly linked to the biological phenomenon that involve the inflammatory and carcinogenic processes, suggesting that the inhibition of GJIC is involved in nongenotoxic cancer induction and tumor promotion [22]. It is reported that disruption of GJIC by hydrogen peroxide is protected by vitamin C [21], however, synthetic antioxidants such as Trolox has no ability to prevent hydrogen peroxide–mediated inhibition of GJIC [22], indicating the role of vitamin C on GJIC appears to be related to a different mechanism, such as the inhibition of signal transduction [22–24]. Therefore, it can be hypothesized that the chemopreventive effects of vitamin C in carcinogenesis may be associated with the protective effects of vitamin C against epigenetic mechanisms, such as inflammation and inhibition of GJIC, as well as to antioxidant activities [24] (**Figure 2**).

2. Epigenetics

Theory of epigenetics takes attention at processes involving in multistage carcinogenesis, which promotional phase of carcinogenesis is a consequence of epigenetic events involving inflammation and the inhibition of GJIC, mediating by ROS [25, 26].

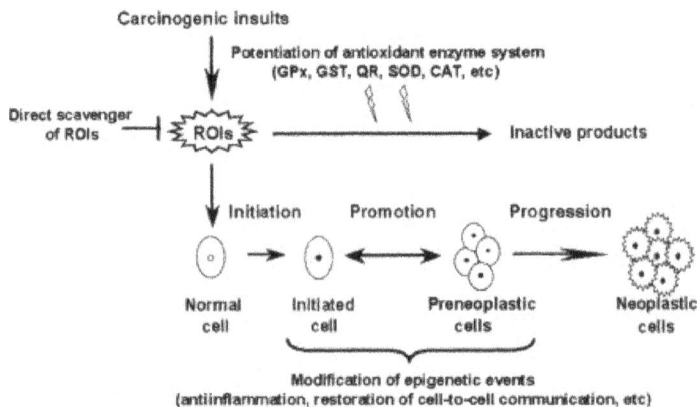

Figure 2.
Possible chemopreventive mechanisms of vitamin C in carcinogenesis. ROIs, reactive oxygen intermediates; GPx, glutathione peroxidase; GST, glutathione S-transferase; QR, quinone oxidoreductase; SOD, superoxide dismutase; CAT, catalase [24].

2.1 DNA demethylation and family of TET dioxygenases

It is known that cancers are caused by genetic alterations. These are genetic changes in oncogenes, such as mutations, deletions, amplifications, rearrangements and translocations, and changes in tumor suppressor or microRNA (miRNA) genes, that are involved in multistep carcinogenesis, and accumulate with tumor progression [27]. Nowadays, it is understood that classical genetics cannot explain all the properties of cancer alone, thus epigenetic abnormalities are also involved in tumorigenesis in addition to genetic alterations [27]. Thus, the epigenome means interface of environment and the genome. Epigenetic modifications catalyzed by certain enzymes and cofactors are the key to understanding molecular connections between the epigenome and the environment. Epigenetics is explained as heritable changes in gene expression that are not caused by DNA sequence alterations [28], including epigenetic alterations such as DNA methylation and histone modifications, that differ between cancer and normal cells [28]. The epigenome reflects the interface of a dynamic environment and the genome. Known epigenetic events include covalent modifications on nucleotides and histones, chromatin remodeling, and non-coding RNAs, which collectively constitute the epigenome [29].

Cytosine hydroxymethylation (5hmC) means of demethylating DNA and activating genes, which is a DNA modification associated with transcriptional silencing [30]. Methylation at the C5 position of cytosine (5-methylcytosine, 5mC) is the major modification of DNA and plays important roles in regulating transcription and cellular identity [30]. Methylation at the C5 position of cytosine (5-methylcytosine, 5mC) is the major and best-characterized epigenetic mark of mammalian DNA. However, 5hmC was generally regarded as oxidatively damaged cytosine in genome and may be replaced by DNA-repair mechanism [30, 31]. Methylation of DNA can change the functional state of regulatory regions, but it does not change the base pairing of cytosine, presenting epigenetic mark and is functionally involved in many forms of stable epigenetic repression [30, 31]. In honeybees DNA methylation patterns on cell and organismal fate is exemplified, in which differential DNA methylation determines whether the bee will be a worker or a queen [31, 32]. DNA methylation has also fundamental choices, such as gene silencing that leads to genomic imprinting, suppression of transposable elements, and the establishment of stable cellular identities [30–32], demonstrating that cellular reprogramming by nuclear transfer, cell fusion, and induced pluripotency can radically alter differentiated cellular states [33]. Epigenetic regulation by DNA methylation provides exciting insights into why reprogramming of cell fates is possible, showing that cytosines in mammalian cells can be hydroxymethylated to 5hmC (5-hydroxymethylcytosine). DNA methylation patterns are frequently observed in disease, particularly in cancer, including methylation of CGI promoters for tumor suppressor genes [34], supporting the recurring mutations in tumors, and thus providing insight into many aspects of biology and medicine. Factors that regulate methylation have been linked to human disease and contribute to malignances still remains largely unknown. The breakthrough in understanding the presence of 5hmC in the genome came from studies on a gene family that is known as ten-eleven translocation (TET) [35].

A group of enzymes termed methylcytosine dioxygenase ten-eleven translocation (TET, including TET1, TET2 and TET3) are identified to catalyze the hydroxylation of 5mC to 5-hydroxymethylcytosine (5hmC) [35, 36]. TET1 is located on human chromosome 10q21.3, TET2 on chromosome 4q24, and TET3 on chromosome 2p13.1. TETs further oxidize 5hmC to 5-formylcytosine (5fC) and 5-carboxylcytosine (5caC) [37]. TETs belong to the Fe^{2+} and 2OG-dependent dioxygenase superfamily. All TET proteins contain a catalytic domain which binds to Fe^{2+} and 2-oxoglutarate (2OG) to mediate oxidation of 5mC to 5hmC in DNA [37]. Subsequent experiments confirmed that TETs are Fe^{2+} and 2OG-dependent dioxygenases, and it is demonstrated that the catalytic activity of TETs is dependent on Fe^{2+} and 2OG. 2OG is also a critical intermediate metabolite of the Krebs cycle, and like collagen P4H, the catalytic activity of TET dioxygenases is indeed dependent on Fe^{2+} and 2OG [37]. All TET proteins contain a C-terminal catalytic domain that consists of a cysteine-rich region and a double-stranded β-helix fold characteristic of the Fe(II)- and 2-oxoglutarate (2-OG)-dependent dioxygenase superfamily [38]. These enzymes require Fe(II) as a cofactor metal and 2-OG as a cosubstrate to catalyze their reactions [38]. Demethylation of 5-methylcytosine (5mC) to 5-hydroxymethyl cytosine (5hmC) is shown to be mediated by TET proteins [38]. TET1 is identified as a Fe2+ and 2OG-dependent enzyme that converts 5mC to 5hmC. Thus, 5hmC, 5fC and 5caC have been proposed as demethylation intermediates. So far, our knowledge of TETs-mediated DNA demethylation is the following: the Fe^{2+} and 2OG-dependent TETs consecutively oxide 5mC to 5hmC, then to 5fC and 5caC which can eventually be removed from the genome and substituted by unmodified C, thus completing the process of DNA demethylation. Although it involves multiple steps, the TET-mediated oxidation in combination with base excision repair constitutes the most important and consistent pathway responsible for the active demethylation of DNA. As being a DNA demethylation intermediate, 5hmC also serves as an epigenetic mark with unique regulatory functions [38–40].

2.2 Epigenetic alterations in cancer

It is demonstrated that tumor cells undergo various epigenetic modifications, DNA hypermethylation, that could lead to an imbalance in regulation of apoptotic genes, that is attributed as one of the important factors in the progression and treatment of cancer [38–40]. There are also small non-coding miRNAs are reported as epigenetic regulators, recently, regulating gene expression through posttranscriptional silencing of target genes.

2.2.1 miRNAs

In some cases, the prognosis and the progression of cancer are associated with changes in the expression of miRNAs during tumorigenesis. miRNAs lead to cleave the mRNA or inhibit translation, depending on the sequence complementarity between the miRNA and its target [41]. DNA methylation is a well-defined epigenetic mark. In humans DNA methylation occurs at cytosine residues in cytosine-guanine (CpG) dinucleotides and is controlled by enzymes called DNA methyltransferases (DNMTs), including DNMT1, DNMT3A and DNMT3B [42]. DNA hypermethylation by CpG promoter inactivating transcriptional tumor suppressor genes is known as one of the alterations that contributes to tumorigenesis in cancer cells [42]. In addition, cancer cells also undergoes hypomethylation at tissue-specific repetitive sequences, while these regions are heavily hypermethylated in normal cells [43], contributing DNA hypomethylation to tumorigenesis by causing chromosomal instability or the reactivation of transposable elements [44, 45]. On the other hand, endogenous repeat element-driven activation of the oncogenic tyrosine kinase, CSF1R, is suggested that impaired epigenetic control and the subsequent transcriptional derepression of repeat elements may play roles in tumorigenesis. It is demonstrated that oncogenes can be activated by the derepression of endogenous repeats, in addition to genetic and epigenetic modifications, suggesting that activation of normally silenced genes by promoter DNA hypomethylation is involved in tumorigenesis (**Figure 2**) [45]. miRNAs are also thought to play a role in tumorigenesis by modulating tumor suppressor genes or oncogenes, as epigenetic regulators of DNA methylation and histone modifications [46], contributing to tumorigenesis by controlling various biological processes, such as proliferation, differentiation, and apoptosis through regulation of or interactions with oncogenes or tumor suppressor genes, acting either as an oncogene or tumor suppressor gene depending on their target genes [47]. Up-regulation of miRNAs targeting tumor suppressor genes by overexpression, amplification, or epigenetic derepression might function as oncogenes inhibiting the activity of an anti-oncogenic pathway. By contrast, the genetic mutation, deletion or epigenetic silencing of a tumor suppressor miRNA that normally represses expression of oncogenes might result in derepression of oncogenes, thereby gain of oncogenic function. miRNAs have roles in the development of tumors, are also implicated in tumor progression by affecting migration, adhesion, and invasion of cancer cells [45–47].

2.2.2 Histone modifications

The roles of epigenetics in the development and progression of tumors is well established. A little is known about the epigenetic activation of cancer-associated genes, although it is focused on the epigenetic inactivation of tumor suppressor genes during tumorigenesis and DNA hypomethylation of some genes. Histone modifications are also important epigenetic marks that are involved in chromatin structure and gene expression, in addition to DNA methylation, including covalent modifications of histone tail residues acting in DNA packaging and regulating transcriptional machinery to coding sequences [48]. These histone modifications occur at histone residues, such as lysine, arginine and serine (methylated, acetylated and phosphorylated) [48, 49]. Gene expression may be regulated by interactions between multiple histone modifications [49], for example, in a growing body. In contrast, aberrant histone modifications, in addition to DNA methylation, are recognized as important epigenetic changes during tumorigenesis [50]. Tumor-suppressor genes are enriched with active histone marks in normal cells, while the transcriptional silencing of those genes in cancer cells is reported to be associated with a loss of active histone marks, and are shown to be common features of human cancer cells [50].

Recently, it is reported that overexpression of cancer-promoting genes in cancer is associated with the loss of repressive histone modifications, suggesting overexpression of oncogenes or cancer-promoting genes in tumors may contribute to tumorigenesis during epigenetic derepression (**Figure 3**). Epigenetically regulated genes may be promising therapeutic targets and biomarkers during tumor initiation. In fact, epigenetic mechanisms involved in the regulation of cancer-associated genes possible epigenetic therapies targeting epigenetically dysregulated genes are contributed to the improvement of patient outcomes [51, 52] (**Figure 3**).

Vitamin C on DNA demethylation is also reported [53]. It is found that vitamin C could directly enhance the catalytic activity of TET dioxygenases by uniquely interacting with the C-terminal catalytic domain of TETs, modulating the epigenetic control of genome activity [53]. Vitamin C acting as a cofactor in DNA demethylation catalyzed by TETs, deficiency of this vitamin, results in disruption of the methylation-demethylation dynamics of DNA and histone, which can contribute to phenotypic alterations or diseases [53].

3. Vitamin C as an epigenetic agent

Vitamin C is an water-soluble micronutrient that exists as ascorbate anion under physiological pH conditions. It is well established that ascorbate is an essential cofactor in various enzymatic reactions and, also and antioxidant and free radical

CARCINOGENESIS

DNA hipermetilation
Histone methylations
Histone modifications
Up-regulated miRNAs

DNA hypometilation
Histone methylations
Histone modifications
Down-regulated miRNAs

Epigenetic Inactivation | Epigenetic Derepression

Tumor supressor genes

Oncogenes or cancer promoting genes

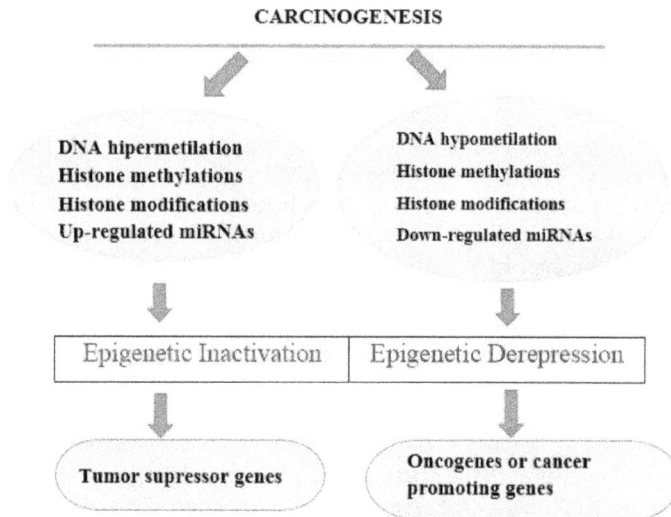

Figure 3.
Epigenetic regulation of cancer-associated genes. Modified from [45].

scavenger. It assists collagen P4H to complete the hydroxylation and thus prevents scurvy; overall, it is required to maintain a number of Fe^{2+} and 2OG-dependent dioxygenases in their fully active forms [53]. In recent years, it is identified that a number of novel Fe^{2+} and 2OG-dependent dioxygenases catalyze the hydroxylation of methylated nucleic acids (DNA and RNA) and methylated histones. Methylation of DNA and histone are the major epigenetic hallmarks in the mammalian genome [53, 54]. It has been shown that some of these dioxygenases require ascorbate as a cofactor to start DNA demethylation and histone demethylation processes [53, 54]. Vitamin C is suggested to effect the genome activity via regulating epigenomic processes [53, 54]. It serves as a cofactor TET dioxygenases that catalyze the oxidation of 5mC into 5hmC. Vitamin C also required for the JmjC domain-containing histone demethylases, acting as a cofactor for histone demethylation [53, 54]. Thus, by participating in the demethylation of both DNA and histones, vitamin C appears to be a mediator between the genome and environment. These findings demonstrate an unknown function of vitamin C in regulating the epigenome, which needs a re-evaluation of the functions of vitamin C in human health and diseases.

Life styles, such as smoking has effects on vitamin C availability. Smoking has also been shown to reduce vitamin C levels in the plasma, strongly [55]. Deficiency of vitamin C results in digestive diseases such as ulcerative colitis, Crohn's disease, chronic gastrointestinal and kidney diseases [55, 56]. Insufficient vitamin C intake during pregnancy, affect the embryonic development due to the changing in the catalytic activity of TETs, resulting in certain types of developmental defects, such as neural tube defect (NTD), and also increases the risk of gastroschisis which is a congenital defect of the abdominal wall [57].

Vitamin C has a long controversial history as a treatment for cancer. As a cofactor for TETs, it enhances the catalytic activity of TETs in cancer cells, and acts in the reprogramming of cancer cells by enhancing the activity of TETs. In cancerous cells, the TET-mediated DNA active demethylation appears to be downregulated [34–40], and a low level of 5hmC is identified as a novel epigenetic hallmark of cancer [34–40]. Mutations in TETs lead to the loss of 5hmC in cancer [34–40]. If vitamin C is deficient, enzymatic activity of TETs are adversely effected, resulting in 5hmC reduction. Studies have demonstrated higher

EPIGENETIC THERAPHY

miRNA targeted agents, DNMTi,
HMTi, HDACi, HDMi

Tumor suppressor genes	Tumor suppressor genes Drug response genes	mi RNAs regulating stem cell genes

- Decrease of tumor growth
- Induction of apoptosis
- Supression of metastasis

Resensitation to chemoteraphy

Decrease of self renewal

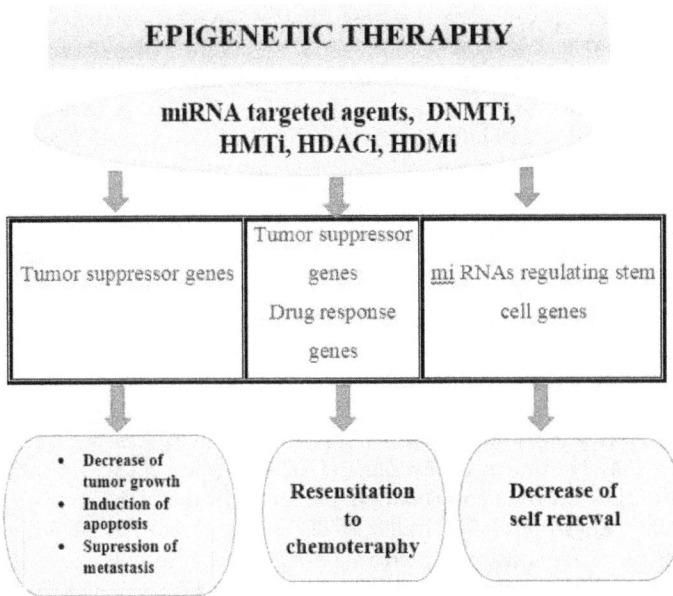

Figure 4.
Epigenetic therapy. Modified from [45].

incidence of scurvy in cancer patients [58, 59]. An association between ascorbate transporters and cancer is also indicated [58, 59]. These findings suggest that vitamin C plays a critical role in the demethylation of DNA and histone, serving as a cofactor for TET, thus, deficiency of this vital vitamin may disrupt the methylation-demethylation dynamics of DNA and histone, and may contribute to phenotypic alterations in different cells along the developmental stages and aging, cancer and other diseases [58, 59], (**Figure 4**).

3.1 Vitamin C and epigenetic regulation in cancer

Studies over the past decades has declared that normal epigenetic regulation is disrupted during tumorigenesis [60], indicating that DNA methylation is the most common event in carcinogenesis [60]. DNA methylation of CpGs in promoters can result in silencing of various genes, including tumor suppressors [61], as a consequence, disrupted gene expression in cancerous cells also can cause alterations in the methylation of lysine or arginine amino acids on histone tails [62].

Since hypermethylation of promoters of tumor suppressor genes been identified as one of the important factors supporting cancer development, demethylation agents are become the main focus of molecular-targeted therapeutics. Vitamin C is shown to play a central role in the conversion of 5mC to 5hmC by enhancing the catalytic activity of TET dioxygenases [56, 57], indicating, vitamin C is an important factor in reducing the risk of promoter hypermethylation and supporting the maintenance of the 5hmC state that plays a major role in the epigenetic regulation [56–58].

3.1.1 Vitamin C and DNA demethylation

Requirement for vitamin C as an additional cofactor for dioxygenases indicated a potential role for this reducing vitamin in TET-mediated DNA demethylation, suggesting that vitamin C has the capacity to modify DNA methylation in human cells [56], for instance, it is known that epigenetic reprogramming occurs during the embryonic development, which involves DNA demethylation and re-methylation. Vitamin C is found to cause DNA demethylation of nearly 2000 genes in embryonic stem cells [56]. These results indicate that vitamin C can be involved in the DNA demethylation process, however, it is unclear whether vitamin C participates directly in DNA demethylation or it is mediated by the enhanced catalytic activity of TETs by vitamin C. DNA hypermethylation is found at promoters of tumor suppressors and developmental genes, typically, whereas CpG sites in other gene regions are hypo-methylated in cancerous cells. 5hmC is lower in a most of cancers, due to a loss of TET activity [56–58], as a result of inactivating mutations, down-regulation of TET gene expression, or insufficient supply of TET co-factors. In embryonic stem cells, in which vitamin C addition is shown to promote DNA demethylation through increased TET activity [63], supports the idea of sufficient vitamin C is important for maintenance of normal 5hmC levels. High-dose vitamin C is also shown to compensate for the loss of TET proteins [64]. Histone modifications and the expression of the genes that regulate these modifications are frequently disrupted in cancer by mutations, translocations/amplifications, or deletions [65]. The loss of TET activity caused by DNA hyper-methylation *in vitro* demonstrated increased methylation at tumor suppressor gene promoters [64, 65], suggesting that vitamin C is a cofactor for TET dioxygenases in the conversion of 5mC to 5hmC, thus modulating DNA demethylation [65] (**Figure 5**).

3.1.2 Vitamin C and histone demethylation

The basic unit of eukaryotic chromatin is composed of a short length of DNA wrapped around an octamer that consists of 2 copies of each histone (H2A, H2B, H3 and H4), called nucleosome. Histones are substrates for post-translational modifications (PTM). PTMs on histones include methylation, acetylation, phosphorylation, and others [66]. The dynamic PTMs in the histone regulate genome stability, gene transcription and chromatin structure. Methylation at lysine and arginine amino acids is epigenetic modification in histones. Histone methylation is a key component

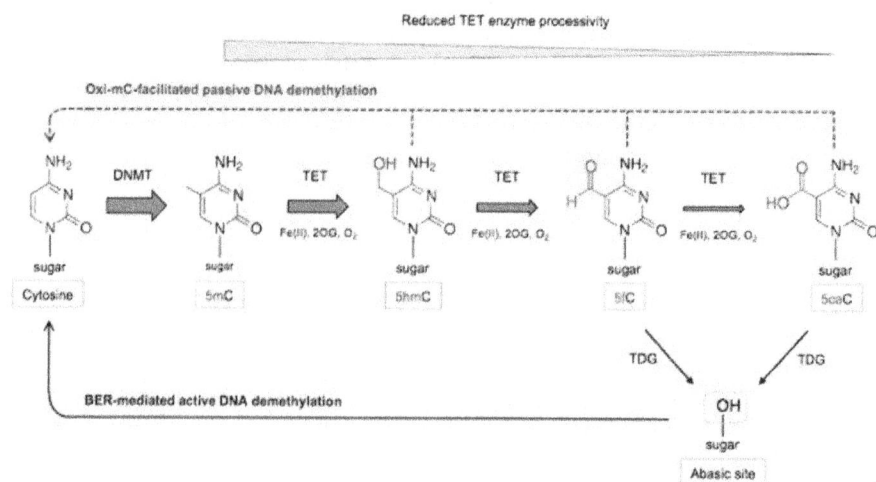

Figure 5.
The role of TET proteins in DNA demethylation [65].

in the epigenome along with DNA methylation. The most studied methylation occurs on histone H3 at lysine (K) 4 (H3K4), H3K9, H3K27, H3K36, H3K79, H4K20 and on histone H3 at arginine (R) 2 (H3R2), H3R8, H3R17, H3R26, H4R3. The methyl donor in histone methylation is S-adenosylmethionine (SAM), the same donor for DNA methylation. JmjC domain histone demethylase 1 (JHDM1) was purified, which specifically demethylates H3K36 in the presence of Fe^{2+} and 2OG [67]. After a long time, 20 proteins that belong to the JmjC domain have been discovered to have the catalytic capacity to demethylate histones [67]. It is now known that the JmjC domain-containing histone demethylases, like TETs, also belong to the Fe^{2+} and 2OG dioxygenase superfamily, that demethylate mono-, di-, and trimethylated histone lysine residues [68], indicating that vitamin C is required for optimal catalytic activity of JHDM1 [68]. It appears to be important in the late phase of reprogramming from terminally differentiated cells, which is also involved in cell differentiation, such as T cell maturation, indicating the role of vitamin C in demethylation of DNA and histone in T cell maturation [69]. The role of vitamin C in histone demethylation is only examined in *in vitro* assays, deducing that vitamin C can be a cofactor for the JmjC domain-containing histone demethylase family, thus modulating histone demethylation in a similar way as it does on DNA demethylation [68–70].

3.1.3 Vitamin C and the loss of 5hmC in cancer

In contrast to the high level of 5hmC in embryos, cancer cells have very low or undetectable 5hmC. It is reported that the loss of 5hmC is a novel epigenetic hallmark of most, especially in certain types of human cancer [56–58]. The loss of 5hmC is resulted in disruption in DNA methylation-demethylation processes leading to malignant transformation [56–58]. Mutations in TETs or a decreased expression of TETs are also attributed to the loss of 5hmC in cancerous cells [56–58], requiring further studies to determine whether there is a local vitamin C deficiency in cancer cells. Linus Pauling proposed the treatment of cancer patients with intravenous vitamin C, in 1970s [71], followed by other investigators, such as Mutlu Demiray in Turkey, nowadays [72]. Epigenetic modulation of vitamin C in the gene activity might shed a new light on this issue. As a cofactor for TETs, vitamin C is found to maximize the catalytic activity of the TETs in cancer cells. In the light of these findings, it can be suggested that the rebuilding the 5hmC content can offer a potential treatment for certain cancers.

3.1.4 Vitamin C in epigenetic treatment of cancer

Vitamin C is a safe and well-tolerated dietary supplement that is utilized in patient care and in treating cancer. Tumor cells are known to be resistant to programmed cell death. It is considered to induce E-cadherin expression in sensitizing tumor cells towards apoptosis [73], as increased expression of E-cadherin is declared to sensitize cancerous cells to cell death [74]. Thus, reactivation of E-cadherin seems to be an important target for epigenetic therapy in cancer. Researchers observed an increase in the expression of E-cadherin by a combination of 5-AZA + vitamin C [75]. They reported an increase in E-cadherin expression by treatment with vitamin C and highlighted the role of vitamin C as an epigenetic player, opening a window for vitamin C-enhanced, TET-dependent conversion of 5mC to 5hmC [75]. Vitamin C consumption may also increase the activity of other epigenetic regulators such as histone demethylases, for which new drugs are currently being developed. Based on the effects of vitamin C on the methylation of DNA and histones, epigenetic regulation has implications in all cancers.

4. Conclusion

Vitamin C is an essential compound with functions far beyond scurvy prevention. As an important mediator between genome and environment, it participates in the demethylation of DNA and histones, epigenome. Genetic and environmental factors that influence the synthesis, absorption, transportation and metabolism of vitamin C could have significant consequences for health and disease by regulating the epigenetic control of genome activity. Cancer is driven by epigenetic modifications along with genetic changes. Vitamin C activates the TET enzymes which are responsible for the removal of methyl groups for DNA and histones, regulating DNA demethylation as an essential cofactor for TET dioxygenases, and regulating histone demethylation as an essential cofactor for Jmjc domain-containing histone demethylases. Deficiency in vitamin C contributes to different diseases, resulting of failure to maintain the catalytic activity of TET dioxygenases and Jmjc domain-containing histone demethylases. Diet and lifestyle are known to affect the level of vitamin C in the human body, dramatically. Deficiency in vitamin C is seen in cancer patients frequently, thus, adequate dietary vitamin C in these patients is needed increasingly, who have mutations in epigenetic regulators. This novel epigenetic function of vitamin C needs to become recognized by the general public. Future studies needs be done for greater understanding of vitamin C impact upon TET and the epigenome which have medicinal relevance in cancer and other diseases.

Author details

Fadime Eryılmaz Pehlivan
Department of Biology, Faculty of Science, Botany Section, Istanbul University, Istanbul, Turkey

*Address all correspondence to: eryilmazfadime@gmail.com

IntechOpen

References

[1] Grzybowski A, Pietrzak K. Albert Szent-Györgyi (1893-1986): The scientist who discovered vitamin C. Clinics in Dermatology. 2013;**31**(3):327-331

[2] Noctor G, Foyer CH. Ascorbate and glutathione: Keeping active oxygen under control. Annual Review of Plant Physiology and Plant Molecular Biology. 1998;**49**:249-279

[3] Arrigoni O, de Tullio MC. The role of ascorbic acid in cell metabolism: Between gene-directed functions and unpredictable chemical reactions. Journal of Plant Physiology. 2000;**157**:481-488

[4] Ki WL, Hyong JL, Young-Joon S, Chang Yong L. Vitamin C and cancer chemoprevention: Reappraisal. The American Journal of Clinical Nutrition. 2003;**78**:1074-1078

[5] Blokhina O, Virolainen E, Fagerstedt KV. Antioxidants, oxidative damage and oxygen deprivation stress. Annals of Botany. 2003;**91**:179-194

[6] Du J, Cullen JJ. Ascorbic acid: Chemistry, biology and the treatment of cancer. Biochimica Et Biophysica Acta-Reviews on Cancer. 2012;**1826**:443-457

[7] Matzinger M, Fischhuber K, Heiss EH. Activation of Nrf2 signaling by natural products—Can it alleviate diabetes? Biotechnology Advances. 2018;**36**:1738-1767

[8] Atanasov AG, Yeung AWK, Banach M. Natural products for targeted therapy in precision medicine. Biotechnology Advances. 2018;**36**:1559-1800

[9] Azmi AS, Sarkar FH, Hadi SM. Pro-oxidant activity of dietary chemopreventive agents: An under-appreciated anti-cancer property. F1000Res. 2013;**2**:135

[10] Padayatty SJ, Katz A, Wang Y, Eck P, Kwon O, Lee JH. Vitamin C as an antioxidant: Evaluation of its role in disease prevention. Journal of the American College of Nutrition. 2003;**22**:18-35

[11] Mikirova NA. The effect of high dose IV vitamin C on plasma antioxidant capacity and level of oxidative stress in cancer patients and healthy subjects. Journal of Orthomolecular Medicine. 2007;**22**:3

[12] Lee KW, Lee HJ, Surh YJ, Lee CY. Vitamin C and cancer chemoprevention: Reappraisal. The American Journal of Clinical Nutrition. 2003;**78**:1074-1080

[13] Schwartz JL. The dual roles of nutrients as antioxidants and prooxidants: Their effects on tumor cell growth. The Journal of Nutrition. 1996;**126**:1221-1227

[14] Wiseman H, Halliwell B. Damage to DNA by reactive oxygen and nitrogen species: Role in inflammatory disease and progression to cancer. Biochemical Genetics. 1996;**313**:17-29

[15] Feiz HR, Mobarhan S. Does vitamin C intake slow the progression of gastric cancer in *Helicobacter pylori*-infected populations? Nutrition Reviews. 2002;**60**:34-36

[16] Drake IM, Davies MJ, Mapstone NP. Ascorbic acid may protect against human gastric cancer by scavenging mucosal oxygen radicals. Carcinogenesis. 1996;**17**:559-562

[17] Wu F, Tyml K, Wilson JX. Ascorbate inhibits iNOS expression in endotoxin- and IFN gamma-stimulated rat skeletal muscle endothelial cells. FEBS Letters. 2002;**520**:122-126

[18] Bowie AG, O'Neill LAJ. Vitamin C inhibits NF-κB activation by TNF via

the activation of p38 mitogen-activated protein kinase. Journal of Immunology. 2000;**165**:7180-7188

[19] Kumar MN, Gilula NB. The gap junction communication channel. Cell. 1996;**84**:381-388

[20] Trosko JE. Commentary: Is the concept of "tumor promotion" a useful paradigm? Molecular Carcinogenesis. 2001;**30**:131-137

[21] Upharm BL, Kang KS, Cho HY, Trosko JE. Hydrogen peroxide inhibits gap junctional intercellular communication in glutathione sufficient but not glutathione deficient cells. Carcinogenesis. 1997;**18**:37-42

[22] Rosenkranz HS, Pollack N, Cunningham AR. Exploring the relationship between the inhibition of gap junctional intercellular communication and other biological phenomena. Carcinogenesis. 2000;**21**:1007-1011

[23] Wu CT, Morris JR. Genes, genetics, and epigenetics: A correspondence. Science. 2001;**293**:1103-1105

[24] Surh Y-J. Molecular mechanisms of chemopreventive effects of selected dietary and medicinal phenolic substances. Mutation Research. 1999;**428**:305-327

[25] Young JI, Züchner S, Wang G. Regulation of the epigenome by vitamin C. Annual Review of Nutrition. 2015;**35**:545-564

[26] Gillberg L, Ørskov AD, Liu M, Laurine BS, Harsløfa PAJ, Grønbæk K. Vitamin C—A new player in regulation of the cancer epigenome. Seminars in Cancer Biology. 2018;**51**:59-67

[27] Li D, Guo B, Wu H, Tan L, Lu Q. TET family of dioxygenases: Crucial roles and underlying mechanisms. Cytogenetic and Genome Research. 2015;**146**:171-180

[28] Kohli RM, Zhang Y. TET enzymes, TDG and the dynamics of DNA demethylation. Nature. 2013;**502**:472-479

[29] Minor EA, Court BL, Young JI, Wang G. Ascorbate induces ten-eleven translocation (Tet) methylcytosine dioxygenase-mediated generation of 5-hydroxymethylcytosine. The Journal of Biological Chemistry. 2013;**288**:13669-13674

[30] Bhutani N, Burns DM, Blau HM. DNA demethylation dynamics. Cell. 2011;**146**:866-872

[31] Schübeler D. Function and information content of DNA methylation. Nature. 2015;**517**:321-326

[32] Kucharski R, Maleszka J, Foret S, Maleszka R. Nutritional control of reproductive status in honeybees via DNA methylation. Science. 2008;**319**:1827-1830

[33] Yamanaka S, Blau HM. Nuclear reprogramming to a pluripotent state by three approaches. Nature. 2010;**465**:704-712

[34] Hore TA. Modulating epigenetic memory through vitamins and TET: Implications for regenerative medicine and cancer treatment. Epigenomics. 2017;**9**:863-871

[35] Abdel-Wahab O, Mullally A, Hedvat C, Garcia-Manero G, Patel J. Genetic characterization of TET1, TET2, and TET3 alterations in myeloid malignancies. Blood. 2009;**114**:144-147

[36] Dong C, Zhang H, Xu C, Arrowsmith CH, Min J. Structure and function of dioxygenases in histone demethylation and DNA/RNA demethylation. IUCrJ. 2014;**1**:540-549

[37] Rasmussen KD, Helin K. Role of TET enzymes in DNA methylation, development, and cancer. Genes & Development. 2016;**30**:733-750

[38] Huang Y, Rao A. Connections between TET proteins and aberrant DNA modification in cancer. Trends in Genetics. 2014;**30**:464-474

[39] Wu H, Zhang Y. Mechanisms and functions of Tet protein-mediated 5-methylcytosine oxidation. Genes & Development. 2011;**25**:2436-2452

[40] Yin X, Xu Y. Structure and function of TET enzymes. Advances in Experimental Medicine and Biology. 2016;**945**:275-302

[41] Wong KY, Huang X, Chim CS. DNA methylation of microRNA genes in multiple myeloma. Carcinogenesis. 2012;**33**:1629-1638

[42] Iorio MV, Piovan C, Croce CM. Interplay between microRNAs and the epigenetic machinery: An intricate network. Biochimica et Biophysica Acta (BBA)-Gene Regulatory Mechanisms. 2010;**10**:694-701

[43] Fabbri M, Calin GA. Epigenetics and miRNAs in human cancer. Advances in Genetics. 2010;**70**:87-99

[44] Wang S, Wu W, Claret FX. Mutual regulation of microRNAs and DNA methylation in human cancers. Epigenetics. 2017;**12**:187-197

[45] Kwon MJ, Shin YK. Epigenetic regulation of cancer-associated genes in ovarian cancer. International Journal of Molecular Sciences. 2011;**12**:983-1008

[46] Fuks F. DNA methylation and histone modifications: Teaming up to silence genes. Current Opinion in Genetics & Development. 2005;**15**:490-495

[47] Ventura A, Jacks T. MicroRNAs and cancer: Short RNAs go a long way. Cell. 2009;**136**(4):586-591

[48] Cedar H, Bergman Y. Linking DNA methylation and histone modification:

Patterns and paradigms. Nature Reviews Genetics. 2009;**10**:295-304

[49] Herranz M, Esteller M. DNA methylation and histone modifications in patients with cancer: Potential prognostic and therapeutic targets. Methods in Molecular Biology. 2007;**361**:25-62

[50] Ellis L, Atadja PW, Johnstone RW. Epigenetics in cancer: Targeting chromatin modifications. Molecular Cancer Therapeutics. 2009;**8**(6):1409-1420

[51] Smith LT, Otterson GA, Plass C. Unraveling the epigenetic code of cancer for therapy. Trends in Genetics. 2007;**23**(9):449-456

[52] Humeniuk R, Mishra PJ, Bertino JR, Banerjee D. Molecular targets for epigenetic therapy of cancer. Current Pharmaceutical Biotechnology. 2009;**10**(2):161-165

[53] Guz J, Oliński R. The role of vitamin C in epigenetic regulation. Postępy Higieny i Medycyny Doświadczalnej. 2017;**71**(1):747-760

[54] Camarena V, Wang G. The epigenetic role of vitamin C in health and disease. Cellular and Molecular Life Sciences. 2016;**73**(8):1645-1658

[55] Schectman G, Byrd JC, Gruchow HW. The influence of smoking on vitamin C status in adults. American Journal of Public Health. 1989;**79**(2):158-162

[56] Anupam Aditi MD, David Y, Graham MD. Vitamin C, gastritis, and gastric disease: A historical review and update. Digestive Diseases and Sciences. 2012;**57**(10):1-22

[57] Smithells RW, Sheppard S, Schorah CJ. Vitamin deficiencies and neural tube defects. Archives of Disease in Childhood. 1976;**51**(12):944-950

[58] Fain O, Mathieu E, Thomas M. Scurvy in patients with cancer. BMJ. 1998;**316**(7145):1661-1662

[59] Mayland CR, Bennett MI, Allan K. Vitamin C deficiency in cancer patients. Palliative Medicine. 2005;**19**: 1-5. DOI: 10.1191/0269216305pm970oa

[60] Wilting RH, Dannenberg JH. Epigenetic mechanisms in tumorigenesis, tumor cell heterogeneity and drug resistance. Drug Resistance Updates. 2012;**15**(1-2):21-38

[61] Baylin SB. DNA methylation and gene silencing in cancer. Nature Clinical Practice Oncology. 2005;**2**:4-11

[62] Vaissière T, Sawan C, Herceg Z. Epigenetic interplay between histone modifications and DNA methylation in gene silencing. Mutation Research/ Reviews in Mutation Research. 2008;**659**(1-2):40-48

[63] Ferrarelli LK. Epigenetic regulation by vitamin C. Science Signaling. 2013;**6**:113. DOI: 10.1126/ scisignal.2004337

[64] Ramalho-Santos M, Laird D, Blaschke K, Ebata KT, Karimi MM, Zepeda-Martínez JA, et al. Vitamin C induces Tet-dependent DNA demethylation in ESCs to promote a blastocyst-like state. Nature. 2013;**500**(7461):222-226. DOI: 10.1038/nature12362

[65] An J, Rao A, Ko M. TET family dioxygenases and DNA demethylation in stem cells and cancers. Experimental & Molecular Medicine. 2017;**49**:323

[66] Hou H, Yu H. Structural insights into histone lysine demethylation. Current Opinion in Structural Biology. 2010;**20**(6):739-748

[67] Hu Q , Baeg GH. Role of epigenome in tumorigenesis and drug resistance. Food and Chemical Toxicology. 2017;**109**:663-668

[68] Monfort A, Wutz A. Breathing-in epigenetic change with vitamin C. EMBO Reports. 2013;**14**(4):337-346

[69] Manning J, Mitchell B, Appadurai DA, Shakya A, Pierce LJ, Wang H, et al. Vitamin C promotes maturation of T-cells. Antioxidants & Redox Signaling. 2013;**19**(17):2054-2067

[70] Kuiper C, Vissers MC. Ascorbate as a co-factor for Fe- and 2-oxoglutarate dependent dioxygenases: Physiological activity in tumor growth and progression. Frontiers in Oncology. 2014;**4**:359

[71] Cameron E, Pauling L. Cancer and vitamin C: A discussion of the nature, causes, prevention, and treatment of Cancer with special reference to the value of vitamin C, updated and expanded. In: Cancer and Vitamin C. Philedelphia, PA: Camino Books; 1993

[72] http://www.mutludemiray.com/ vitamin-c-nin-kanserle-iliskisi.html

[73] Pećina-Šlaus N. Tumor suppressor gene E-cadherin and its role in normal and malignant cells. Cancer Cell International. 2003;**3**:17

[74] Capra J, Eskelinen S. Correlation between E-cadherin interactions, survivin expression, and apoptosis in MDCK and ts-Src MDCK cell culture models. Laboratory Investigation. 2017;**97**:1453-1470

[75] Sajadian SO, Tripura C, Samani FS, Ruoss M, Dooley S, Baharvand H, et al. Vitamin C enhances epigenetic modifications induced by 5-azacytidine and cell cycle arrest in the hepatocellular carcinoma cell lines HLE and Huh7. Clinical Epigenetics. 2016;**8**:46

Chapter 4

Vitamin C, Aged Skin, Skin Health

Philippe Humbert, Loriane Louvrier, Philippe Saas
and Céline Viennet

Abstract

Vitamin C is an essential nutriment for humans. Vitamin C is known for its antioxidant potential. Vitamin C acts as a potent water-soluble antioxidant in biological fluids. Thus, topical vitamin C will not only reduce the risks of development of photoaging but also could reduce the risk of carcinogenesis. In addition to its antioxidant properties, vitamin C is essential for collagen synthesis. Vitamin C stimulates or restores several mechanisms which are either deficient or disturbed. Topical application of vitamin C partially restores the anatomical structure of the epidermal-dermal junction in young skin. A clinical trial confirmed for the first time that topical application of 5% vitamin C over a period of 6 months significantly improves the clinical appearance of photodamaged skin when compared to the vehicle alone. In inflammatory skin diseases, that is, atopic dermatitis and psoriasis, vitamin C levels into the dermis are reduced. Moreover, a randomized double-blind comparative study conducted in patients with Bateman purpura showed a significant improvement that vitamin C is probably one of the main topical anti-aging agents. In addition, the use of photo-protective sunscreen after UV irradiation prevents the decrease of acid ascorbic dermis concentration. Indeed, the ingestion of vitamin C has different benefits on skin such as wound healing, cutaneous aging, and prevention of skin cancer.

Keywords: vitamin C, dermatoporosis, Bateman purpura, wound healing, collagen, healing

1. Introduction

Vitamin C is an essential nutriment for humans. The name "ascorbic" is derived from ascorbutic (scorbutus: scurvy). This disease was described by the ancient Egyptians, Greeks, and Romans and during the crusade in the thirteenth century. Its deficiency is responsible for scurvy. It is characterized by altered functions of the connective tissue, such as perifollicular hemorrhages and defective healing. In the eighteenth century, Lind, a British naval surgeon, found that ingestion of lemon juice cured this disease [1]. Since the discovery in the 1930s that vitamin C is the antiscorbutic factor, much works have been undertaken to elucidate its mechanisms of action. L-enantiomer is the most bioactive form of ascorbic acid. It is natural compound produced in most plants and animals; however, humans lack the enzyme L-gulono-gamma-lactone oxidase, which is necessary to produce it. Vitamin C is an alpha-ketolactone [2]. Humans are one of the few species that require dietary supplementation of ascorbic acid for survival.

2. Photo aging

Bearing in mind the mechanisms of photo damages, the beneficial role of vitamin C or L-ascorbic acid has been raised. Thus, vitamin C is the major aqueous phase antioxidant agent in humans. In addition to its antioxidant properties, vitamin C is essential for collagen synthesis. Vitamin C stimulates or restores several mechanisms, which are either deficient or disturbed.

The environmental damage, particularly ultraviolet irradiation (UV), induces human skin aging or photoaging. Indeed, 80% of facial aging is believed to be due to chronic sun exposure [3]. In addition, it is well known that UV from the sun has deleterious effects in human skin, including sunburn, immune suppression, and cancer [4, 5]. UVA radiation (λ, 320–400 nm) and infrared and even visible light may produce skin damage like a UVB radiation (λ, 290–320 nm) [6]. However, erythema, aging, and carcinogenesis are still assigned mainly to UVB and UVA [7]. The UV irradiation like other several exogenous (pollution, stress, and smoking) and endogenous factors (normal metabolic processes) invokes a complex sequence of specific molecular responses that damage skin connective tissue, particularly production of reactive oxygen species (ROS).

The photochronological generation of ROS is a primary mechanism by which UV irradiation initiates molecular responses in human skin. These ROS contain superoxide anion, peroxide, and singlet oxygen. Further, some evidence suggests that free radicals induce alterations in gene expression pathways, which in turn contribute to the degradation of collagen and the accumulation of elastin emblematic of photoaged skin [8]. The ROS activate cell surface receptors such as epidermal growth factor, interleukin (IL1), insulin, keratinocytes growth factor, and tumor necrosis factor (TNF α) *in vivo*.

The UV irradiation actives NADPH oxidase that is responsible for generating hydrogen peroxide too. The NADPH oxidase catalyzes the reduction of molecular oxygen to superoxide anion [9]. The hydrogen peroxide is distinct from photochemical generation of ROS, which occurs only during UV exposure and abates following UV exposure. Hydrogen peroxide is less damaging to cells and can be converted to other ROS, such as hydroxyl radical and singlet oxygen [9]. It is known that UV irradiation significantly upregulates the synthesis of several types of collagen-degrading enzymes known as matrix metalloproteinases (MMPs). The UV irradiation activates protein kinase-mediated signaling pathways. The activated kinase upregulate expression and functional nuclear transcription factor, AP-1. It follows the stimulation of transcription of genes for matrix-degrading enzymes, such as metalloproteinase (MMP) 1 (collagenase), MMP-3 (stromelysin 1), and MMP-9 (92 Kd gelatinase) [10, 11]. The UV-induced MMP-1 initiates cleavage of fibrillar collagen (type I and III in skin) at a single site within its central triple helix. This collagen can be further degraded by elevated levels of MMP-3 and MMP-9 [12]. The MMP induced by UV degrades skin collagen, disrupting the structural architecture of dermis. In addition, UV irradiation alters the synthesis of collagen, primarily through downregulation of type I and type III pro-collagen gene expression [13]. Collagen I is the most abundant protein in skin, and type I and type III collagen fibrils provide strength and resiliency to skin. Photoaged skin contains an abundance of degraded, disorganized collagen fibrils and has a reduced production of type I and type III pro-collagen. Fibroblasts are elongated and collapsed [14].

Alterations in elastic fibers are so strongly associated with photoaged skin that "elastosis," an accumulation of amorphous elastin material, is considered pathognomonic of photoaged skin. The UV irradiation induces a thickening and coiling of elastic fibers in the papillary dermis.

In addition, other alterations of many important structural components of the dermal extra cellular matrix are observed such as modifications in the structure and composition of anchoring fibrins, proteoglycans, and glycoaminoglycans [15, 16].

It is well known that skin pigment provides a significant degree of protection against actinic damage. Therefore, in lighter skin color after UV irradiation, erythema is greater and tanning is less in darker skin pigment. In addition, the effect of skin color on UV-induced MMP-1 gene expression and formation of thymine dimmers were investigated in human skin *in vivo*. UVB and UVA exposure resulted in substantial induction of MMP-1 mRNA and formation of thymine dimmers in lightly pigmented subjects group. In contrast, twice the average exposure of the lightly pigmented group produced only modest MMP-1 mRNA induction or DNA damage in the darkly pigmented group [9].

Skin photoaged induces variable changes that affect the sun-exposed areas (face, neck, forearms, and dorsum of hands), such as deep wrinkles not erased by stretching, roughness, yellow hue, leathery appearance [17–19], atrophy and laxity, a focal depigmentation (guttate hypomelanosis) and/or hyperpigmentary changes (bronzing, ephelides, and actinic lentigines), purpura, telangiectasia or venous lakes, comedoes (Favre-Raccouchot disease), and a neoformation on photoaged skin (benign and premalignant melanocytic, epithelial neoformation, and basal and squamous cell carcinomas). These aspects change among individuals. Degree of photoaging is great in individuals who have outdoor life styles, live in sunny climates, and are lightly phototype [20]. Although wrinkles can be found almost everywhere in an aged skin, they develop preferentially on photo-exposed areas and are thus largely visible by others. Little is known even about the pathogenesis of wrinkles [21]. In fact, skin laxity, due to decreased volumes of the inner-layers, and loss of intra-dermal tensional strength are presumed to play a major role in wrinkling [22]. In addition, it is claimed that the decreased dermal area compared to epidermis area, due to the disappearance of the dermal papillae and the weaker attachment by an altered dermo-epidermal junction (DEJ), is one of the physical prerequisites that allow wrinkle formation. The *stratum corneum* of the wrinkle is said to be thickened by an accumulation of corneocytes, forming a real horny plug at the bottom of the wrinkle. The spinous layer of a wrinkle was shown to be thinner at the base than at the flanks [23], and fewer keratohyaline granules are present in the wrinkle base as compared to the flanks. Furthermore, filaggrin decreases at the bottom of wrinkle. DEJ has also been shown to contain less collagen IV and VII at the bottom of the wrinkle than in its flanks or in the surrounding skin. Concerning dermis, Tsuji et al. earlier have reported [24] a decrease of actinic damage at the bottom of wrinkle compared to its sides or adjacent skin; this shows an accumulation of highly damaged elastotic material. In addition, oxytalan fibers have nevertheless been reported to be sparse or absent at the bottom of the wrinkle, and chondroitin sulphate GAGs have also been shown to be reduced under the wrinkles.

3. Vitamin C (ascorbate): the main potent antioxidant vitamin

Vitamin is known for its antioxidant potential. Vitamin C acts as a potent water-soluble antioxidant in biological fluids by scavenging physiologically relevant reactive oxygen species and reactive nitrogen species. It is the main water soluble nonenzymatic antioxidant in aqueous compartments. It does not absorb light in the UVB and UVA range, but it exerts its effects by neutralizing oxygen free radicals (superoxide, hydroxyl and water soluble peroxyl radical, singlet oxygen, and hypochlorous acid) [25], as well as nonradical species such as hypochlorous acid, ozone, singlet oxygen, nitrosating species, nitroxide, and peroxynitrite. It is acknowledged that vitamin C works in extracellular fluids, as blood plasma and respiratory tract lining fluid. Low vitamin C blood levels are associated with enhanced oxidative stress, specifically reported in smoking and inflammatory diseases as rheumatoid arthritis and adult respiratory distress syndrome. In addition, vitamin C attenuates oxidative damage by generating small antioxidant molecules, such as α-tocopherol, glutathione, urate, and β-carotene. α-tocopherol may be regenerated at the expense of vitamin C at the interface between a lipid membrane and aqueous phase and therefore is protected from peroxidation. The mode of interaction between vitamin C and α-tocopherol related to antioxidant capacities is complex, depending on the *in vivo* context.

Vitamin C is an effective donor antioxidant, and its active form is ascorbate (reduced form). It loses electrons in antioxidant reactions and becomes oxidized to the short lived ascorbyl radical. The ascorbyl radical is relatively unreactive due to resonance stabilization of the unpaired electron; it readily dismutes ascorbate and dehydroascorbic acid (DHA) by NADH-dependent semidehydroascorbate reductase. DHA can be recycled back to ascorbate by NADPH-dependent thioredoxin reductase or glutathione-dependent DHA reductase.

Vitamin C plays a crucial role in various hydroxylation reactions. It acts as a cofactor in the enzymatic synthesis of collagen, carnitine, catecholamine, and neurohormones. As a cofactor for hydroxylase and oxygenase metalloenzymes, vitamin C is thought to work by reducing the active metal site, resulting in reactivation of the metal-enzyme complex or by acting as a co-substrate involved in the reduction of molecular oxygen. Vitamin C uses iron and copper as cofactors, and it enhances intestinal iron absorption. The reduction of iron by vitamin C is involved in the enhanced dietary absorption of nonheme iron.

It should be noted, however, that ascorbate might also act as a pro-oxidant. Hydroxyl radicals and lipid alkoxyl radicals can be generated in the presence of reduced metal ions with hydrogen peroxide and lipid hydroperoxides. These interactions involve redox reactions called the Fenton chemistry. Although this Fenton chemistry occurs easily *in vitro*, its relevance *in vivo* is still a matter of controversy. The effects of ascorbate depend on the *in vivo* availability of catalytic metal ions. In healthy individuals, iron is largely sequestered by iron binding proteins such as transferrin, ferritin, and ceruloplasmin. Instead of this, during tissue injury, metal ions may be released and could interact with ascorbate. Therefore, the ability of vitamin C to modulate metal ion metabolism will be discussed.

Fenton reaction mediated by vitamin C. (1) Vitamin C reduces ferric ions (Fe^{3+}) to ferrous ions (Fe^{2+}). (2) Ferrous ion reacts with oxygen to produce superoxide. (3) Dismutation of superoxide leads to hydrogen peroxide. (4) Hydrogen peroxide reacts with ferrous ions to form hydroxyl radicals.

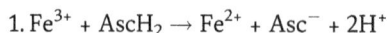

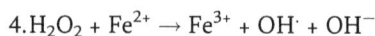

1. $Fe^{3+} + AscH_2 \rightarrow Fe^{2+} + Asc^- + 2H^+$

2. $Fe^{2+} + O_2 \rightarrow Fe^{3+} + O_2{}^-$

3. $2O_2{}^- + 2H^+ \rightarrow H_2O_2 + O_2$

4. $H_2O_2 + Fe^{2+} \rightarrow Fe^{3+} + OH^{\cdot} + OH^-$

4. Effects of vitamin C on the skin

In addition to its antioxidants properties, vitamin C and some water-soluble derivatives are essential for collagen biosynthesis [26]. In fibroblasts cultures, vitamin C stimulates collagen biosynthesis. Its absence results in structurally unstable collagen, which is not secreted from cells at a normal rate [27]. The role of vitamin C in the hydroxylation of collagen molecules is well characterized [28]. In fact, ascorbic acid is an essential cofactor for the enzyme, prolyl, and lysyl hydroxylases, catalyzing the synthesis of hydroxyproline and hydroxylysine in collagen. Hydroxyproline acts to stabilize the collagen triple helix. Its absence results in structurally unstable collagen, which is not secreted from cells at a normal rate. Hydroxylysine is necessary for cross-linking one collagen molecule to another, providing tissue strength [29, 30]. This occurs concurrently with a decrease in elastin production, and the elastin protein is often overproduced in response to photodamage [31]. Vitamin C also increases the proliferation rate of fibroblasts, a capacity that is decreased with age [32].

Specifically, vitamin C has been shown to stabilize collagen mRNA, thus increasing collagen protein synthesis for repair of the damaged skin [33]. Further, vitamin C stimulates DNA repair in cultured fibroblasts [34].

The epidermal dermal junction of aged and photoaged skin is flattened due to the loss of rete ridges and the disappearance of papillary projections. In extremely aged skin, papillae virtually disappear, and the junction with the atrophic epidermis is a straight line versus undulations in younger skin with a thin papillary dermis along with a loss of capillaries. The corneocytes in aged skin become larger as a result of decreased epidermal turn over.

Topical application of vitamin C partially restores the anatomical structure of the epidermal-dermal junction in young skin and increases the number of nutritive capillary loops in the papillary dermis close to the epidermal tissue in the aged skin of postmenopausal women. The increase of the density of papillae after vitamin C treatment is linked to a restructuring of the papillary dermis, as the top of the newly formed papillae and the capillaries are localized above the average height of the plane basal layer seen predominantly in aged skin. Moreover, new blood vessels are formed during the treatment with vitamin C. These vessels show a normal anatomical structure in confocal microscopical examination and are apparently integrated in a healthy vascular architecture. The mechanism, by which VC restores dermal papillae, is unknown. These results suggest that topical vitamin C may have important antiaging effects in correcting the structural and functional losses associated with skin aging [35].

Vitamin C protects keratinocytes from the damage produced by ultraviolet A [36]. Vitamin C transport proteins are increased in keratinocytes in response to UV light, suggesting an increased need for vitamin C uptake for adequate protection [37, 38]. UV light decreases vitamin C content of skin, an effect that is dependent on the intensity and duration of UV exposure [39–42]. In cultured keratinocytes, the addition of vitamin C reduces UV-related DNA damage and lipid peroxidation, limits the release of pro-inflammatory cytokines, and protects against apoptosis [36–42]. Vitamin C also modulates redox-sensitive cell signaling in cultured skin cells and consequently increases cell survival following UV exposure [43, 44].

4.1 Systemic administration of vitamin C

Vitamin C is important to the body, including skin. Therefore, as little as 10 mg per day is necessary to prevent scurvy. In the United States, the current recommended daily allowance (RDA) is 60 mg. This rate increases for smokers and pregnant or lasting women. The supplementation of vitamin C has beneficial effects in human body, such as wound healing, prevention of cancer, cataract prevention, and atherosclerosis, and enhances immune mechanisms [45]. The saturated body store of vitamin C is approximately 20 mg/kg of body weight, corresponding to 0.9 mg/dl of a plasma ascorbate level [46]. The half-life of this vitamin is 10–20 days and is dependent on plasma levels [46]. In case of total deficiency, the organism reserves depleted approximately after 4 weeks. The cutaneous levels of vitamin C were well determined [47]. Ascorbate is distributed in all layers of the skin. However, in the epidermis, level of vitamin C is more than five times the level in the dermis (3.8 µmol/g skin versus 0.72 µmol/g skin). Those differences reflect the role of antioxidants in the epidermis and dermis.

The effect of aging on antioxidant capacity in murine skin is well known [39]. With age, the antioxidant activity in skin decreases, and UV enhances this phenomenon [48].

Vitamin C from oral supplements appears to accumulate in the skin, reducing the basal content of malonaldehyde and an oxidation residue of lipids. Oral supplementation with vitamin C (500 mg/day) has shown no evidence for an effect of the vitamin C on UVR-induced oxidative stress [49]. Fuchs and Kern have shown that a very high supplementation dose of 3 grams vitamin C daily increases significantly the minimal erythema dose [50].

Two observational studies found that higher intakes of vitamin C from the diet were associated with better skin appearance, with notable decreases in skin wrinkling [51, 52].

Many studies have proved the synergistic interaction of vitamin C and vitamin E (D-alpha-tocopherol) in antioxidant defense and in decrease of sunburn reaction. UV induces a chain reaction of lipid peroxidation in membranes rich in polyunsaturated fatty acids. Vitamin E protects the membranes by stopping the propagation reactions of lipid peroxyl radicals, while ascorbic acid simultaneously recycles alpha-tocopherol.

Vitamin C is also required to form competent barrier lipids in the epidermis [53] by stimulating the synthesis of ceramides. It has also been shown to stimulate the barrier function of the endothelial cells [54].

Besides, Vitamin C supplementation has been reported to inhibit skin, nerve, and lung and kidney carcinogenesis. Vitamin C has been shown to inhibit tumor cell carcinogen-induced DNA damage [55].

Its beneficial activity as a photo protectant [56] and anticancer agent [57] has been demonstrated by dietary supplementation in humans and in animal species even in those that can synthesize the vitamin.

5. Local vitamin C

Aging causes a decline in vitamin C content in both the epidermis and dermis [58, 59], and it has been demonstrated a correlation between decreasing ascorbic acid dermis level and age: vitamin C concentration is higher by young women [59]. Leveque et al. have shown that there is a direct relationship between iron and vitamin C concentrations in the human dermis and aging. Therefore, the level of vitamin C decreases linearly with age [60]. Moreover, the vitamin C concentration skin decreases after exposure to UV lights [61]. In addition, it was demonstrated that the use of photoprotective sunscreen after UV irradiation prevents the decrease of acid ascorbic dermis concentration [61]. Since sun exposure induces reduction in vitamin C levels in the dermis, it appears interesting to bring to the skin topical vitamin C in such cases. Thus, topical vitamin C will reduce the risks of development of photoaging but also could reduce the risk of carcinogenesis. Animal experiments have demonstrated the UV photoprotective effect of topically applied vitamin C. Darr et al. [62] showed, in porcine model, that 10% aqueous vitamin C that was applied several times to the animal's skin, which was irradiated with 400 mJ/cm^2, reduced UV-induced skin photoinjury (erythema and sunburn cell formation) via its antioxidant potential. In the hairless mouse model, Bissett et al. [63] showed that a solution of ascorbic acid (5%) applied 2 hours before exposure and reduced chronic skin damage from UVB and UVA.

An *in vitro* study realized by Boxman et al. [64] used the heat shock protein HSP 27 as a sensitive marker of skin irritation or cellular stress in reconstructed skin. Stress (exposure to sodium lauryl sulphate or UV light) generated, in reconstructed skin, relocalization of HSP 27 from the cytoplasm to the cell nucleus in the absence of vitamin C and no relocalization in presence of vitamin C. In this study, they suggested that vitamin C may control the response to exterior stress in reconstructed skin. Topical application of ascorbate exerts a protective effect on the inflammatory response to sun exposure and on the UV immunosuppression. Perricone [65] has shown that ascorbyl palmitate, applied after UV burning, reduced redness 50% sooner than areas on the same patient which were left untreated. In mice's study, Steenvoorden [66] demonstrated that a single topical application of vitamin C at 0.5–5 μmol/cm^2 protected the skin against UV-induced systemic immunosuppression. Many studies have proved the synergistic interaction of vitamin C and vitamin E (D-alpha-tocopherol) in antioxidant defense [67] and in the decrease of sunburn reaction.

UV induces a chain reaction of lipid peroxidation in membranes rich in polyunsaturated fatty acids. Alpha-tocopherol protects the membranes by stopping the propagation reactions of lipid peroxyl radicals, while ascorbic acid simultaneously recycles alpha-tocopherol.

The study led by Lin et al. [68] evaluated if a combination of topical vitamins C and E is better for UV protection of skin than an equivalent concentration of topical vitamin C or E alone. Results showed that the individual ingredients were associated with a twofold increase in the antioxidant protection factor compared with vehicle control, while the association of L-ascorbic acid and alpha-tocopherol produced a fourfold increase in the antioxidant protection factor. Another study [69, 70] showed that the addition of ferulic acid, an antioxidant, improves the chemical stability of vitamin C and E and produces a further doubling of the photoprotective effect (from four to eightfold).

On the basis of *in vitro* and *in vivo* studies, it has been postulated but also proven that vitamin C could be used topically for prevention and correction of skin aging. Lots of results show an improvement of the clinical appearance of photoaged skin with vitamin C topical application. Photoprotective properties of topically applied vitamin C have thus placed this molecule as a potential candidate for use in prevention and treatment of skin aging.

Vitamin C can reverse the ROS (reactive oxygen species)-induced skin damage. An *in vivo* antioxidant capacity of vitamin C was demonstrated after a week of topical application. Topical application of vitamin C before sun exposure limits the UV-induced oxidant stress.

The histological pattern of aging skin is the flattening of the dermal-epidermal junction which is accompanied by a reduction of the density of papillae in the dermal-epidermal transition zone [71]. In *in vivo* study, Sauermann et al. [72] determined the effect of a topical cream containing 3% vitamin C against the excipient alone using daily applications for 4 months on the forearm of 33 women. They showed with the confocal laser scanning microscopy method that topical vitamin C may have therapeutically effects for partial corrections of the regressive structural changes with the aging process notably to enhance the density of dermal papillae, perhaps through the mechanism of angiogenesis.

Kameyama et al. [73] have studied the effect of a stable derivative of ascorbic acid and magnesium-ascorbyl-2-phosphate (MAP) on melanogenesis *in vitro* and *in vivo*. They showed that the topical application of MAP is effective in lightening the skin of some patients with hyperpigmentation disorders (melasma or solar lentigines). This vitamin C derivative suppresses melanin formation by influence of tyrosinase and melanoma cells.

Few studies [52, 74] have shown the role of ascorbate in the formation of *stratum corneum* barrier lipids. It seems that ascorbate normalizes epidermal lipid profiles, in particular glucosphingolipids and ceramides.

Lots of studies [75–77] have shown an improvement of the clinical appearance of photoaged skin with vitamin C topical application. Humbert et al. [78] evaluated the clinical effects and the modifications of skin relief and structure over a 6 month period of use of a cream containing 5% vitamin C on photoaged skin. They found modifications of the skin relief (wrinkles, roughness, and skin elasticity) and the skin ultrastructure, suggesting a positive influence of topical vitamin C on parameters characteristic for sun-induced skin aging.

A double-blind randomized trial has been performed to evaluate the clinical effects and the modifications of skin relief and structure over a 6-month period of use of a cream containing 5% vitamin C on photoaged skin. The aim of the present study was to evaluate the effect on the dermal cells of vitamin C administrated by topical application on the skin of normal human volunteers by measuring the steady state level of the mRNAs of procollagens, their post-translational processing enzymes, the fibrillar structures-associated proteoglycan, decorin, the main components of the elastic fibers, elastin, and fibrillins 1 and 2, and the enzymes involved in the degradation of these matrix components. Such investigations performed on small biopsies were made possible by the use of quantitative RT-PCR controlled by original newly created internal standards of synthetic RNA. The mRNAs were measured by RT-PCR made quantitative by simultaneous amplification of internal standards of synthetic RNA.

Clinical assessments included evaluation at the beginning and after 3 and 6 months of daily treatment performed by the investigator and volunteer self-assessment. Skin relief parameters were determined on silicone rubber replicas performed at the same time points. Cutaneous biopsies were obtained at the end of the trial and investigated using immunohistochemistry and electron microscopy. Clinical examination by a dermatologist as well as self-assessment by the volunteers disclosed a significant improvement, in terms of "aging score," of the vitamin C–treated side versus control. A highly significant increase in the density of skin microrelief and a decrease of the deep furrows were demonstrated.

The results of this clinical trial confirmed for the first time based on a randomized controlled blind study that topical application of 5% vitamin C over a period of 6 months significantly improves the clinical appearance of photodamaged skin when compared to the vehicle alone. Significant favorable modifications of skin relief were induced, leading to the reappearance of an isotropic surface pattern. Significant reduction in small and coarse wrinkles and improvement in the overall aspect of the skin assessed by the volunteers were observed after 6 months of daily treatment. But such evolution of skin relief was already noticed 3 months after the beginning of treatment.

For the first time, this study disclosed fibroblast effects of topical vitamin C, with ultrastructural evidence of the elastic tissue to be repaired. Indeed 10 patients accepted a skin biopsy on their forearms, each being treated with either the vitamin

C emulsion or the placebo. Therefore, at the termination of the treatment, two 5-mm punch biopsies up to the hypodermis were collected under local anesthesia at the site of the topical application. One biopsy was used for classical morphology, and electron microscopy reported elsewhere (manuscript in preparation). The second biopsy was used for measurement of the mRNAs.

The mRNA of collagen type I and type III was increased to a similar extent by vitamin C. The mRNA of three posttranslational enzymes, procollagen N-, C-proteinases, and lysyloxidase, was similarly increased. The mRNA of decorin was also stimulated, but elastin, fibrillin 1 and 2, MMP1, 2, and 9 were not modified by the vitamin. The stimulating activity of topical vitamin C was most conspicuous in the women with the lowest dietary intake of the vitamin. The results demonstrate that vitamin C penetrates up to the dermis and further indicate that collagen synthesis is not maximal in postmenopausal women and can be increased.

These data clearly showed that topically applied vitamin C can have a beneficial effect for treatment of skin aging, the mechanism of action being related to an activation of the collagen metabolism, an in activation of a dermal synthesis of elastic fibers.

An example of skin relief improvement.

Same area in a patient at T0, after 3 months of treatment, and after 6 months of treatment.

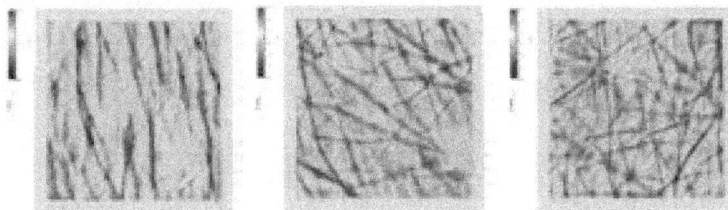

Campos et al. [79] compared the effects of vitamin C and its derivatives on the skin after 2 and 4 week period daily applications of topical formulation containing ascorbic acid or derivatives (magnesium ascorbyl phosphate (MAP) and ascorbyl tetraisopalmitate (ATIP)). Results showed that vitamin C derivatives did not present the same effects of ascorbic acid on human skin. They obtained an enhancement in transepidermal water loss (TEWL) with ascorbic acid which due to enhancement of the epidermal cell renewal process on human skin and an increase of cutaneous hydration in the deeper cell layers with MAP.

6. How to manage topical vitamin C

The stratum corneum is the primary obstacle to efficient vitamin C absorption from external sources [80]. Although concentrations of vitamin C up to 30% have been used for animal studies, maximal absorption was achieved with a 20% vitamin C solution, with higher concentrations showing lower absorption [80]. Topical application of ascorbic acid will cross the epidermis into the underlying dermal layers. A major obstacle for L-ascorbic acid application in topical formulation is its low stability in aqueous media. The stability of vitamin C in topical solutions is a concern, as exposures to air, heat, and/or light may slowly degrade vitamin C. AA is a sensitive compound, which is degraded by oxidation. Several factors can negatively influence ascorbic acid degradation such as high storage temperatures, light, and high pH values. Catalytic amounts of metals, mainly iron, increase the rate of oxidation [81, 82].

The degradation of vitamin C proceeds within days and weeks depending on formulation, packaging, and storage condition. This loss of active substance is often accompanied by brownish discolorations of the formula, including an increased risk of skin incompatibilities and physical instabilities of the formula system itself. The instability of these compounds limits their application in pharmaceutical and cosmetics industries.

To solve this problem of stability, derivatives of vitamin C have been synthesized having an action similar to ascorbic acid but with improved chemical stability. Two derivatives are widely used in cosmetic products: lipophilic ascorbyl palmitate and hydrophilic ascorbyl phosphate salts. These differ in their ability to permeate the skin, as a result of their different hydrolipophilic properties. Many novel compounds as stabilized ascorbyl pentapeptide (SAP), which is much more stable than L-ascorbic acid in water [83], have been recently studied.

Formulation concepts for increase vitamin C stability have also been formulated on anhydrous system [38], in solution [39] or in other system as O/W/O emulsions [84] and micro emulsions with alkyl polyglucoside [85].

Topical application of vitamin C has been shown to elevate significantly cutaneous levels of vitamin E in pigs, and this correlates with protection of the skin from UVB damage as measured by erythema and sunburn cell formation [86]. A combination of both vitamins C and E provided very good protection from a UVB insult. This study - confirms the utility of antioxidants as photoprotectants but suggests the importance of combining the compounds with known sunscreens to maximize photoprotection.

In a study done by Dreher [87], the topical application of combinations of both vitamins or of melatonin with vitamins enhanced the photoprotective response. The better protection was obtained by using the combination of melatonin with both vitamins.

From another hand, no significant protective effect of melatonin or the vitamins when applied alone or in combination was obtained when antioxidants were applied after UV radiation exposure. No improved photoprotective effect was obtained when multiple applications were done [88].

They investigate the effect of the use of high-frequency ultrasound together with coupling gel containing vitamin C and niacinamide as skin lightening agents. Ultrasound radiation enhanced the absorption of skin lightening agents in the stratum corneum in a radiation time-dependent manner. The data suggest that the use of high-frequency ultrasound radiation together with skin lightning gel is effective to reduce hyperpigmentation via enhancing transepidermal transport of vitamin C and niacinamide [89].

7. Bateman purpura

Bateman purpura is a noninflammatory hemorrhagic syndrome characterized by the presence of purpuric eruptions like a petechial or confluent ecchymosed, which was first described by Bateman in 1836. The skin tends to appear thin and wrinkly and almost look flimsy. It results in flat blotches, which start out red and then turn purple, darken a bit, and eventually fade away. It may also occur in the mucous membrane such as mouth and internal organs. The lesions appeared mainly along the outside of the forearm in successive dark purple blotches of an irregular form and various magnitudes. The lesions are also localized to legs, back of the hands, bridge of the nose in subjects wearing glasses, and on the parts of the body where the skin is lax and inelastic [90]. These are flat, irregular, and purple lesions which appear on the skin as one gets older. Confirmed by Rayer in 1839, Unna described in 1895 six cases. All were aged women, and in all except one, the purpura was limited to the extensor surface of the forearm. Senile purpura frequency is around 10% in an elderly population between 70 and 90 years with a female predominance, and it is associated, in 90% of the cases, with pseudoscars. In 2007, Kaya and Saurat proposed the term "Dermatoporosis" to cover different manifestations and implications of the chronic cutaneous insufficiency/fragility syndrome [91]. The clinical manifestations of Dermatoporosis comprise morphological markers of fragility such as skin atrophy, senile purpura and stellate pseudoscar, and functional expression of skin fragility. They result of skin fragility from minor traumas, such as frequent skin laceration, delayed wound healing, and subcutaneous bleeding with the formation of dissecting hematomas leading to large zones of necrosis. Considering the special and pathognomonic patterns of Bateman purpura, such as skin atrophy, stellar scars, hemorrhages, and ecchymosis, it seemed to us that these symptoms are consistent with the diagnosis of localized scurvy. In order to support this assertion, Humbert et al. demonstrated few years ago that aged skin was deficient in vitamin C content [59]. Furthermore, it is noteworthy that the lesions occur on photo-exposed skin such as forearms, face, neck, etc. Sun exposure is known to deplete the skin in antioxidant vitamins, especially vitamin C [39]. A randomized double-blind comparative study (twice daily application of the active vs. the neutral cream) was conducted in patients with Bateman purpura. Clinical examination by experts showed a significant improvement on the vitamin C–treated side compared with the control. Twice-daily application of 5% topical vitamin C led to a clinically apparent improvement of the skin symptoms. These results confirmed the hypothesis of underlying of role vitamin C deficiency in the determinism of Bateman purpura [92, 93].

8. Atopic dermatitis and inflammatory skin diseases

Atopic dermatitis (AD) is defined as a chronic inflammatory cutaneous disease. It is caused by an immune response occurring in a genetically predisposed background. The mechanisms at the origin of AD are not totally elucidated but have three characteristics: Genetic: 50–70% of the patients have a relative who is affected; immunologic: allergy symptoms are frequent during AD; cutaneous: abnormalities of the cutaneous surface.

At cutaneous level, AD is characterized by repeated pruriginous outbreaks of acute eczema affecting mainly the skin folds. Clinical signs vary with age and according to the stage of the disease. It is possible to distinguish two phases: pruriginous outbreaks of acute eczema and remission. The chronic inflammatory reaction of the patient is associated with constitutive abnormalities or induced abnormalities of the cutaneous barrier, with an increase in the Trans Epidermal Water Loss (TEWL), abnormalities of the surface cutaneous lipids, and other clinical signs such as xerosis or dryness resulting from the deterioration of the barrier function. Besides, atopic dermatitis is an inflammatory skin condition, in which the beneficial role of antioxidant molecules can be underlined.

Vitamin C belongs to the defense mechanisms of the organism. As an example, vitamin C acts like a reducer against the free radicals of oxygen (harmful for the organism) released during the inflammatory phenomenon: it deactivates the radicals peroxides and becomes a stable ascorbyl radical. The mobilization of vitamin C during inflammation could explain the fall of its dermic concentration, particularly in atopic subjects [94], as well in psoriatic patients [95]. In theory, if the anti-radicalizing vitamins (A, C, and E vitamins) applied topically arrive at the site of release of the free radicals, their intervention makes the reduction of the harmful effects of these radicals possible.

A clinical study was performed to assess the cutaneous effects of vitamin C applied topically to subjects with light to moderate atopic dermatitis who applied emulsion at 10% of vitamin C *versus* excipient on their forearms during 56 days. The barrier function parameters were assessed: TEWL, hydration, dryness, pH, and relief, and the clinical evolution of the cutaneous lesions was evaluated by SCORAD.

Overall, the clinical scores decreased over the 56 days of the study. Therefore, the eczema lesions tend to regress: SCORAD located on the forearm after and before treatment decreased of 10 and 4 units in the vitamin C and the excipient, respectively. The eczema lesions tended to regress but without significant difference between active and excipient.

The TEWL decreased and all the parameters evolved positively: increase of hydration, decrease of dryness, and improvement of cutaneous relief (unpublished data).

9. Vitamin C wound

There is evidence that topical vitamin C might be beneficial in several unrelated conditions. Topical vitamin C has been reported to improve wound healing [96]. As scurvy progresses, wound healing is impaired due to the loss of mature collagen, which allows wounds to remain open [97]. Skin lesions caused by vitamin C deficiency are remediated by an adequate intake of vitamin C. Studies on the effect of vitamin C supplementation on wound healing have reported somewhat mixed results. Data from laboratory animals and humans show that vitamin C deficiency results in poor wound healing, and vitamin C supplementation in deficient individuals shows significant benefits. Although vitamin C levels appear to increase collagen synthesis and decrease inflammatory responses at the site of the wound, neither vitamin C supplementation [98] nor increased plasma vitamin C status [99] increases wound closure time in otherwise healthy individuals. This suggests that vitamin C may only affect specific facets of the wound healing response. Topical ascorbic acid has not been properly evaluated prior to or during wound healing in humans.

10. Conclusion

Bearing in mind the mechanisms of photodamages and skin protection against aggressions, it should be desirable to take vitamin C orally or to apply it directly to the skin. In addition to its antioxidants properties, vitamin C and some water-soluble derivatives are essential for collagen biosynthesis. Vitamin C is a free radical scavenger by its antioxidant properties. Moreover, Vitamin C is probably one of the main topical anti-aging agents, with clinical proofs of efficacy. In addition, it was demonstrated that the use of photo protective sunscreen after UV irradiation prevents the decrease of acid ascorbic dermis concentration. Indeed, the ingestion of vitamin C has different benefits on skin such as wound healing, cutaneous aging, and prevention of skin cancer.

Author details

Philippe Humbert*, Loriane Louvrier, Philippe Saas and Céline Viennet
University of Bourgogne Franche-Comté, INSERM, EFS BFC, UMR1098,
Interactions Hôte-Greffon-Tumeur/Ingénierie Cellulaire et Génique, Besançon,
France

*Address all correspondence to: philippe.humbert@univ-fcomte.fr

IntechOpen

References

[1] Hodges RE, Baker EM, Hood J, Sauberlich HE, March S. Experimental scurvy in man. The American Journal of Clinical Nutrition. 1969;**22**(5):535-548

[2] Nishikimi M, Fukuyama R, Minoshima S, Shimizu N, Yagi K. Cloning and chromosomal mapping of the human non-functional gene for L-gulono-gamma-lactone oxidase, the enzyme for L-ascorbic acid biosynthesis missing in man. The Journal of Biological Chemistry. 1994;**269**: 13685-13688

[3] Uitto J. Understanding premature skin aging. The New England Journal of Medicine. 1997;**337**(20):1463-1465

[4] Fisher GJ, Wang ZQ, Datta SC, Varani J, Kang S, Voorhees JJ. Pathophysiology of premature skin aging induced by ultraviolet light. The New England Journal of Medicine. 1997; **337**:1419-1429

[5] Cooper KD, Oberhelman L, Hamilton TA, Baadsgaard O, Terhune M, LeVee G, et al. UV exposure reduces immunization rates and promotes tolerance to epicutaneous antigens in humans: Relationship to dose, CD1a-DR+ epidermal macrophage induction, and Langerhans cell depletion. Proceedings of the National Academy of Sciences of the United States of America. 1992;**89**:8497-8501

[6] Petersen MJ, Hansen C, Craig S. Ultraviolet A irradiation stimulates collagenase production in cultured human fibroblasts. The Journal of Investigative Dermatology. 1992;**99**(4): 440-444

[7] Cascinelli N, Krutmann J, Mackie R, Pierotti M, Prota G, Rosso S, et al. European school of oncology advisory report: Sun exposure, UVA lamps and risk of skin cancer. European Journal of Cancer. 1994;**30A**:548-560

[8] Scharffetter-Kochanek K, Brenneisen P, Wenk J, Herrmann G, Ma W, Kuhr L, et al. Photoaging of the skin from phenotype to mechanisms. Experimental Gerontology. 2000;**35**(3): 307-316

[9] Babior BM, Lambeth JD, Nauseef W. The neutrophil NADPH oxidase. Archives of Biochemistry and Biophysics. 2002;**397**:342-344

[10] Karin M, Liu Z-G, Zandi E. AP-1 function and regulation. Current Opinion in Cell Biology. 1997;**9**:240-246

[11] Angel P, Szabowski A, Schorpp-Kistner M. Function and regulation of AP-1 subunits in skin physiology and pathology. Oncogene. 2001;**20**: 2413-2423

[12] Sternlicht MD, Werb Z. How matrix metalloproteinases regulate cell behavior. Annual Review of Cell and Developmental Biology. 2001;**17**: 463-516

[13] Fisher GJ, Datta S, Wang Z, Li XY, Quan T, Chung JH, et al. c-Jun-dependent inhibition of cutaneous procollagen transcription following ultraviolet irradiation is reversed by all-*trans* retinoic acid. The Journal of Clinical Investigation. 2000;**106**: 663-670

[14] Varani J, Schuger L, Dame MK, Leonard C, Fligiel SEG, Kang S, et al. Reduced fibroblast interaction with intact collagen as a mechanism for depressed collagen synthesis in photodamaged skin. The Journal of Investigative Dermatology. 2004;**122**: 1471-1479

[15] Lavker RM. Cutaneous aging: Chronologic versus photoaging. In: Gilchrest B, editor. Photodamage. Vol. 1. Cambridge: Blackwell Science Inc; 1995. pp. 123-135

[16] Yaar M, Gilchrest BA. Skin aging: Postulated mechanisms and consequent changes in structure and function. Clinics in Geriatric Medicine. 2001;**17**: 617-630

[17] Yaar M, Gilchrest BA. Photoageing: Mechanism, prevention and therapy. The British Journal of Dermatology. 2007;**157**(5):874-887

[18] Chung JH, Hanft VN, Kang S. Aging and photoaging. Journal of the American Academy of Dermatology. 2003;**49**(4):690-697

[19] Rabe JH, Mamelak AJ, McElgunn PJ, Morison WL, Sauder DN. Photoaging: Mechanisms and repair. Journal of the American Academy of Dermatology. 2006;**55**(1):1-19

[20] Kligman AM, Kligman LH. Photoaging. In: Freedberg IM, Eisen AZ, Wolff K, et al., editors. Fitzpatrick's Dermatology in General Medicine. 5th ed. Vol. 1. New York: McGraw-Hill; 1999. 1717-1723

[21] Kligman AM, Zheng P, Lavker RM. The anatomy and pathogenesis of wrinkles. The British Journal of Dermatology. 1985;**113**(1):37-42

[22] Piérard GE, Uhoda I, Piérard-Franchimont C. From skin microrelief to wrinkles. An area ripe for investigation. Journal of Cosmetic Dermatology. 2003;**2**:21-28

[23] Contet-Audonneau JL, Jeanmaire C, Pauly G. A histological study of human wrinkle structures: Comparison between sunexposed areas of the face, with or without wrinkles, and sunprotected areas. The British Journal of Dermatology. 1999;**140**(6):1038-1047

[24] Tsuji T, Yorifuji T, Hayashi Y, Hamada T. Light and scanning electron microscopic studies on wrinkles in aged persons' skin. The British Journal of Dermatology. 1986;**114**:329-335

[25] Nakurama T, Pinnell SR, Streilein JW. Antioxidants can reverse the deleterious effects of ultraviolet (UVB) radiation on cutaneous immunity [letter]. The Journal of Investigative Dermatology. 1995;**104**:600

[26] RM L, Zheng PS, Dong G. Aged skin: A study by light, transmission electron and scanning electron microscopy. The Journal of Investigative Dermatology. 1987;**88**:44s-51s

[27] Peterkofsky B. Ascorbate requirement for hydroxylation and secretion of procollagen: Relationship to inhibition of collagen synthesis in scurvy. The American Journal of Clinical Nutrition. 1991;**54**:1135-1140

[28] Murad S, Grove D, Lindberg KA, Reynolds G, Sivarajah A, Pinnell SR. Regulation of collagen synthesis by ascorbic acid S. Proceedings of the National Academy of Sciences of the United States of America. 1981;**78**: 2879-2882

[29] Tajima S, Pinnell SR. Ascorbic acid preferentially enhances type I and III collagen gene transcription in human skin fibroblasts. Journal of Dermatological Science. 1996;**11**:250-253

[30] Davidson JM, LuValle PA, Zoia O, Quaglino D, Giro M. Ascorbate differentially regulates elastin and collagen biosynthesis in vascular smooth muscle cells and skin fibroblasts by pretranslational mechanisms. The Journal of Biological Chemistry. 1997;**272**(1):345-352

[31] Phillips C, Combs S, Pinnell S. Effects of ascorbic acid on proliferation and collagen synthesis in relation to the donor age of human dermal fibroblasts. The Journal of Investigative Dermatology. 1994;**103**:228-232

[32] Geesin J, Darr D, Kaufman R, Murad S, Pinnell S. Ascorbic acid specifically increases type I and type III

procollagen messenger RNA levels in human skin fibroblast. The Journal of Investigative Dermatology. 1988;**90**: 420-424

[33] Duarte T, Cooke M, Jones G. Gene expression profiling reveals new protective roles for vitamin C in human skin cells. Free Radical Biology & Medicine. 2009;**46**(1):78-87

[34] Boyce ST, Supp AP, Swope VB, Warden GD. VC regulates keratinocyte viability,epidermal barrier, and basement membrane *in vitro*,and reduces wound contraction after grafting of cultured skin substitutes. The Journal of Investigative Dermatology. 2002;**118**:565-572

[35] Tebbe B, Wu S, Geilen CC, Eberle J, Kodelja V, Orfanos CE. L-ascorbic acid inhibits UVA-induced lipid peroxidation and secretion of IL-1 alpha and IL-6 in cultured human keratinocytes *in vitro*. The Journal of Investigative Dermatology. 1997;**108**: 302-306

[36] Steiling H, Longet K, Moodycliffe A, Mansourian R, Bertschy E, Smola H, et al. Sodium-dependent vitamin C transporter isoforms in skin: Distribution, kinetics, and effect of UVB-induced oxidative stress. Free Radical Biology & Medicine. 2007; **43**(5):752-762

[37] Kang JS, Kim HN, Kim JE, Mun GH, Kim YS, Cho D, et al. Regulation of UVB-induced IL-8 and MCP-1 production in skin keratinocytes by increasing vitamin C uptake via the redistribution of SVCT-1 from the cytosol to the membrane. The Journal of Investigative Dermatology. 2007;**127**(3): 698-706

[38] Leveque N, Mac-Mary S, Muret P, Makki S, Aubin F, Kantelip J-P, et al. Evaluation of a sunscreen photoprotective effect by ascorbic acid assessment in human dermis using microdialysis and gas chromatography mass spectrometry. Experimental Dermatology. 2005;**14**(3):176-181

[39] Shindo Y, Witt E, Packer L. Antioxidant defense mechanisms in murine epidermis and dermis and their responses to ultraviolet light. The Journal of Investigative Dermatology. 1993;**100**:260-265

[40] Podda M, Traber MG, Weber C, Yan L-J, Packer L. UV-irradiation depletes antioxidants and causes oxidative damage in a model of human skin. Free Radical Biology & Medicine. 1998;**24**(1):55-65

[41] Shindo Y, Witt E, Han D, Packer L. Dose-response effects of acute ultraviolet irradiation on antioxidants and molecular markers of oxidation in murine epidermis and dermis. The Journal of Investigative Dermatology. 1994;**102**(4):470-475

[42] Savini I, D'Angelo I, Ranalli M, Melino G, Avigliano L. Ascorbic acid maintenance in HaCaT revents radical formation and apoptosis by UV-B. Free Radical Biology & Medicine. 1999; **26**(9):1172-1180

[43] Nakamura T, Pinnell S, Darr D, Kurimoto I, Itami S, Yoshikawa K, et al. Vitamin C abrogates the deleterious effects of UVB radiation on cutaneous immunity by a mechanism that does not depend on TNF-alpha. The Journal of Investigative Dermatology. 1997;**109**: 20-24

[44] Shapiro SS, Saliou C. Role of vitamins in skin care. Nutrition. 2001; **17**:839-844

[45] Hornig D. Metabolism and requirements of ascorbic acid in man. South African Medical Journal. 1981; **60**(21):818-823

[46] Shindo Y, Witt E, Hand D, Epstein W, Packer L. Enzymic and non enzymic

antioxidants in epidermis and dermis of human skin. The Journal of Investigative Dermatology. 1994;**102**: 122-124

[47] Lopez-Torres M, Shindo Y, Packer L. Effect of age on antioxidants and molecular markers of oxidative damage in murine epidermis and dermis. The Journal of Investigative Dermatology. 1994;**102**(4):476-480

[48] McArdle F, Rhodes LE, Parslew R, Jack CIA, Friedmann PS, Jackson MJ. UVR-induced oxidative stress in human skin *in vivo*: Effects of oral vitamin C supplementation. Free Radical Biology & Medicine. 2002;**33**(10):1355-1362

[49] Fuchs J, Kern H. Modulation of UV-light-induced skin inflammation by D-alpha-tocopherol and L-ascorbic acid: A clinical study using solar simulated radiation. Free Radical Biology & Medicine. 1998;**25**(9):1006-1012

[50] Cosgrove M, Franco O, Granger S, Murray P, Mayes A. Dietary nutrient intakes and skin-aging appearance among middle-aged American women. The American Journal of Clinical Nutrition. 2007;**86**:1225-1231

[51] Purba MB, Kouris-Blazos A, Wattanapenpaiboon N, Lukito W, Rothenberg EM, Steen BC, et al. Skin wrinkling: Can food make a difference? Journal of the American College of Nutrition. 2001;**20**(1):71-80

[52] Ponec M, Weerheim A, Kempenaar J, Mulder A, Gooris G, Bouwstra J, et al. The formation of competent barrier lipids in reconstructed human epidermis requires the presence of vitamin C. The Journal of Investigative Dermatology. 1997;**109**:348-355

[53] Utoguchi N, Ikeda K, Saeki K, Oka N, Mizuguchi H, Kubo K, et al. Ascorbic acid stimulates barrier function of cultured endothelial cell monolayer.

Journal of Cellular Physiology. 1995;**163**: 393-399

[54] Chen LH, Boissonneault GA, Glauret HP. Vitamin C, vitamin E and cancer (review). Anticancer Research. 1988;**8**:739-748

[55] Darr D, Dunston S, Faust H, Pinnell S. Effectiveness of antioxidants (Vitamin C and E) with and without sunscreens as topical photoprotectants. Acta Dermato-Venereologica. 1996;**76**: 264-268

[56] Pauling L. Effect of ascorbic acid on incidence of spontaneous mammary tumors and UV-light-induced skin tumors in mice. The American Journal of Clinical Nutrition. 1991;**54**(6 suppl): 1252S-1255S

[57] Rhie G, Shin MH, Seo JY, Choi WW, Cho KH, Kim KH, et al. Aging-and photoaging-dependent changes of enzymic and nonenzymic antioxidants in the epidermis and dermis of human skin *in vivo*. The Journal of Investigative Dermatology. 2001;**117**(5):1212-1217

[58] Lévèque N, Muret P, Mary S, Makki S, Kantelip J, Rougier A, et al. Decrease in skin ascorbic acid concentration with age. European Journal of Dermatology. 2001;**12**(4):21-22

[59] Leveque N, Robin S, Makki S, Muret P, Rougier A, Humbert P. Iron and ascorbic acid concentrations in human dermis with regard to age and body sites. Gerontology. 2003;**49**(2):117-122

[60] Levêque N, Mac-Mary S, Muret P, Makki S, Aubin F, Kantelip J-P, et al. Evaluation of sunscreen photoprotective effect by ascorbic acid assessment in human dermis using microdialysis and gas chromatography mass sprectrometry. Experimental Dermatology. 2004;**00**:1-6

[61] Darr D, Combs S, Dunston S, Manning T, Pinnell S. Topical vitamin C

protects porcine skin from ultraviolet radiation-induced damage. British Journal of Dermatology. 2006;**127**: 247-253

[62] Bissett DL, Chatterjee R, Hannon DP. Photoprotective effect of superoxide-scavenging antioxidants against ultraviolet radiation-induced chronic skin damage in the hairless mouse. Photodermatology, Photoimmunology & Photomedicine. 1990;**7**(2):56-62

[63] Boxman IL, Kempenaar J, de Haas E, Ponec M. Induction of HSP27 nuclear immunoreactivity during stress is modulated by vitamin C. Experimental Dermatology. 2002;**11**(6):509-517

[64] Perricone NV. The photoprotective and anti-inflammatory effects of topical ascorbyl palmitate. The Journal of Dermatology. 1993;**1**:5-10

[65] Steenvoorden DP, Beijersbergen van Henegouwen G. Protection against UV-induced systemic immunosuppression in mice by a single topical application of the antioxidant vitamins C and E. International Journal of Radiation Biology. 1999;**75**(6):747-755

[66] Chan AC. Partners in defense, vitamin E and vitamin C. Canadian Journal of Physiology and Pharmacology. 1993;**71**(9):725-731

[67] Lin J-Y, Selim A, Shea C, Grichnik JM, Omar MM, Monteiro-Riviere NA, et al. UV photoprotection by combination topical antioxidants vitamin C and E. Journal of the American Academy of Dermatology. 2003;**48**:866-874

[68] Lin FH, Lin JY, Gupta RD, Tournas JA, Burch JA, Selim MA, et al. Ferulic acid stabilizes a solution of vitamins C and E and doubles its photoprotection of skin. The Journal of Investigative Dermatology. 2005; **125**(4):826-832

[69] Murray JC, Burch JA, Streilein RD, Iannacchione MA, Hall RP, Pinnell SR. A topical antioxidant solution containing vitamins C and E stabilized by ferulic acid provides protection for human skin against damage caused by ultraviolet irradiation. Journal of the American Academy of Dermatology. 2008;**59**:418-425

[70] Sauermann K, Clemann S, Jaspers S, Gambichler T, Altmeyer P, Hoffmann K, et al. Age related changes of human skin investigated with histometric measurements by confocal laser scanning microscopy *in vivo*. Skin Research and Technology. 2002;**8**:52-56

[71] Sauermann K, Jaspers S, Koop U, Wenck H. Topically applied vitamin C increases the density of dermal papillae in aged human skin. BMC Dermatology. 2004;**29**:13

[72] Kameyama K, Sakai C, Kondoh S, Yonemoto K, Nishiyama S, Tagawa M, et al. Inhibitory effect of magnesium L-ascorbyl-2-phosphate (VC-PMG) on melanogenesis *in vitro* and *in vivo*. Journal of the American Academy of Dermatology. 1996;**34**(1):29-33

[73] Uchida Y, Behne M, Quiec D, Elias PM, Holleran WM. Vitamin C stimulates sphingolipid production and markers of barrier formation in submerged human keratinocyte cultures. The Journal of Investigative Dermatology. 2001;**117**(5):1307-1313

[74] Raschke T, Koop U, Düsing HJ, Filbry A, Sauermann K, Jaspers S, et al. Topical activity of ascorbic acid: From *in vitro* optimization to *in vivo* efficacy. Skin Pharmacology and Physiology. 2004;**17**:200-206

[75] Traikovich SS. Use of topical ascorbic acid and its effects on photodamaged skin topography. Archives of Otolaryngology – Head & Neck Surgery. 1999;**125**:1091-1098

[76] Fitzpatrick RE, Rostan EF. Double blind, half-face study comparing topical vitamin C and vehicle for rejuvenation of photodamage. Dermatologic Surgery. 2002;**28**(3):231-236

[77] Humbert P, Haftek M, Creidi P, Lapière C, Nusgens B, Richard A, et al. Topical ascorbic acid on photoaged skin. Clinical, topographical and ultrastructural evaluation: Double-blind study vs. placebo. Experimental Dermatology. 2003;**12**:237-244

[78] Maia Campos PMBG, Gonçalves GMS, Gaspar LR. *In vitro* antioxidant activity and *in vivo* efficacy of topical formulations containing vitamin C and its derivatives studied by non-invasive methods. Skin Research and Technology. 2008;**14**:376-380

[79] Pinnell SR, Yang H, Omar M, Riviere NM, DeBuys HV, Walker LC, et al. Topical L-ascorbic acid: Percutaneous absorption studies. Dermatologic Surgery. 2001;**27**(2): 137-142

[80] Buettner GR, Jurkiewicz BA. Chemistry and biochemistry of ascorbic acid. In: Packer L, editor. Handbook of Antioxidants. Chapter 5. New York: M. Dekker; 1996. pp. 96-100

[81] Austria R, Semenzato A, Bettero A. Stability of vitamin C derivatives in solution and topical formulations. Journal of Pharmaceutical and Biomedical Analysis. 1997;**15**:795-801

[82] Choi HI, Park JI, Kim HJ, Kim DW, Kim SS. A novel L-ascorbic acid and peptide conjugate with increased stability and collagen biosynthesis. BMB Reports. 2009;**42**(11):743-746

[83] Farahmand S, Tajerzadeh H, Farboud ES. Formulation and evaluation of a vitamin C multiple emulsion. Pharmaceutical Development and Technology. 2006;**11**(2):255-261

[84] Pakpayat N, Nielloud F, Fortuné R, Tourne-Peteilh C, Villareal A, Grillo I, et al. Formulation of ascorbic and microemulsion with alkyl polyglycosides. European Journal of Pharmaceutics and Biopharmaceutics. 2009;**72**:444-452

[85] Darr D, Combs S, Dunston S, Manning T, Pinnell S. Topical vitamin C protects porcine skin from ultraviolet radiation-induced damage. British Journal of Dermatology. 1992;**127**: 247-253

[86] Dreher F, Gabard B, Schwindt D, Maibach HI. Topical melatonin in combination with vitamins E and C protects skin from ultraviolet-induced erythema: A human study *in vivo*. British Journal of Dermatology. 1998; **139**:332-339

[87] Dreher F, Denig N, Gabard B, Schwindt DA, Maibach HI. Effects of topical antioxidants on UV-induced erythema formation when administered after exposure. Dermatology. 1999;**198**: 52-55

[88] Hakozaki T, Takiwaki H, Mijamoto K, Sato Y, Arase S. Ultrasound enhanced skin—lightening effects of Vitamin C and Niacinamide. Skin Research and Technology. 2006;**12**:105-113

[89] Tattersall RN, Seville R. Senile purpura. The Quarterly Journal of Medicine. 1950;**19**(74):151-159

[90] Shiozawa S, Tanaka T, Miyahara T, Murai A, Kameyama M. Age-related change in the reducible cross-link of human skin and aorta collagens. Gerontology. 1979;**25**(5):247-254

[91] Kaya G, Saurat J-H. Dermatoporosis: A chronic cutaneous insufficiency/ fragility syndrome. Dermatology. 2007; **215**(4):284-294

[92] Humbert P, Fanian F, Lihoreau T, Jeudy A, Pierard GE. Bateman purpura

(dermatoporosis): A localized scurvy treated by topical vitamin C—double-blind randomized placebo-controlled clinical trial. Journal of the European Academy of Dermatology and Venereology. 2018;**32**:323-328

[93] Humbert P, Louvrier L. Photoexposed skin, skin ageing, Bateman's purpura and local vitamin C deficiency. Journal of the European Academy of Dermatology and Venereology. 2018;**32**(10):e383-e384

[94] Leveque N, Robin S, Muret P, et al. High iron and low ascorbic acid concentrations in the dermis atopic dermatitis patients. Dermatology. 2003;**2007**:261-264

[95] Leveque N, Robin S, Muret P, Mac-Mary S, Makki S, Berthelot A, et al. *In vivo* assessment of iron and ascorbic acid in psoriatic dermis. Acta Dermato-Venereologica. 2004;**8**:2-5

[96] Boyce ST, Supp AP, Swope VB, Warden GD. Vitamin C regulates keratinocyte viability, epidermal barrier, and basement membrane *in vitro*, and reduces wound contraction after grafting of cultured skin substitutes. The Journal of Investigative Dermatology. 2002;**118**(4):565-572

[97] Ross R, Benditt E. Wound healing and collagen formation II. Fine structure in experimental scurvy. The Journal of Cell Biology. 1962;**12**(3):533-551

[98] Silverstein RJ, Landsman AS. The effects of a moderate and high dose of vitamin C on wound healing in a controlled guinea pig model. The Journal of Foot and Ankle Surgery. 1999;**38**(5):333-338

[99] Sørensen LT, Toft BG, Rygaard J, Ladelund S, Paddon M, James T, et al. Effect of smoking, smoking cessation, and nicotine patch on wound dimension, vitamin C, and systemic markers of collagen metabolism. Surgery. 2010;**148**(5):982-990